11   Biometrical Genetics
12   Theories of Plant and Animal
     Breeding
13   Ecological and Evolutionary
     Genetics

THE OPEN UNIVERSITY

Science:
A Second Level Course

S299
GENETICS

Prepared by a Course Team for the Open University

THE OPEN UNIVERSITY PRESS

Course Team

Chairman and General Editor
Steven Rose

Unit Authors
Norman Cohen (*The Open University*)
Terence Crawford-Sidebotham (*University of York*)*
Denis Gartside (*University of Hull*)
David Jones (*University of Hull*)
Steven Rose (*The Open University*)
Derek Smith (*University of Birmingham*)
Mike Tribe (*University of Sussex*)
Robert Whittle (*University of Sussex*)

*Consultant

Editor
Jacqueline Stewart

Other Members
Bob Cordell (*Staff Tutor*)
Mae-Wan Ho*
Jean Holley (*Technician*)
Stephen Hurry
Roger Jones (*BBC*)
Aileen Llewellyn (*Course Assistant*)
Michael MacDonald-Ross (*IET*)
Jean Nunn (*BBC*)
Pat O'Callaghan (*Evaluation*)
Jim Stevenson (*BBC*)
* From January 1976

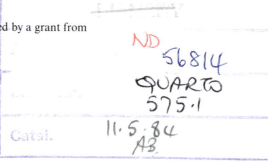

The development of this Course was supported by a grant from the Nuffield Foundation.

The Open University Press,
Walton Hall, Milton Keynes.

First published 1976.

Designed by the Media Development Group of the Open University.

Set by Composition House Ltd, Salisbury, Wiltshire.

Printed in Great Britain by Eyre and Spottiswoode Limited,
at Grosvenor Press, Portsmouth.

ISBN 0 335 04288 0

This text forms part of an Open University Course. The complete list of Units in the Course appears at the end of this text.

For general availability of supporting material referred to in this text please write to the Director of Marketing, The Open University, P.O. Box 81, Walton Hall, Milton Keynes, MK7 6AT.

Further information on Open University Courses may be obtained from the Admissions Office, The Open University, P.O. Box 48, Walton Hall, Milton Keynes, MK7 6AB.

1.1

11 Biometrical Genetics

Contents

List of scientific terms used in Unit 11

Developed in this Unit	Page No.	Developed in this Unit	Page No.
additive component	520	experimental designs	513
backcross generations	520	F_1 hybrids	542
Cavalli analysis	534	gene dispersion	526
components of generation means	520	heterosis	536
		inbreeding depression	536
continuous variation	508	incomplete dominance	518
degree of dominance	520	mid-parent value, m	515
dominance	517	multifactorial inheritance	508
dominance component	520	potence ratio	529
epistasis	534	scaling tests	530
environmental effects	513		

Objectives for Unit 11

After studying this Unit you should be able to:

1   Define, recognize the best definition of, and place in the correct context, the items in the list of scientific terms above.

2   Given the mean values of the parental and F_1 generations for a difference in a single gene, compute the mean value of subsequent generations by deriving m, d and h.
(SAQ 1)

3   Given the mean values of the parental and F_1 generations, calculate the degree of dominance of the parental gene.
(SAQ 1)

4   Given the mean values of the parental and F_1 generations for differences in many genes, calculate the mean value of subsequent generations by deriving m $[d]$ and $[h]$.
(SAQs 2, 3 and 4)

5   Given the mean values for the parental and F_1 generations, calculate the potence ratio.
(SAQs 2 and 4)

6   Demonstrate the importance of scaling tests by calculating the value of the A-statistic and also its standard error.
(SAQ 4)

7   Using biometrical terms, define heterosis and explain its two possible causes.
(SAQ 5)

8   Discuss the practical advantages and disadvantages of breeding heterotic F_1 hybrids in crop-plant species.

Study guide for Unit 11

In this Unit we describe continuous variation by means of mathematical models and statistical parameters. Unit 11 progresses slowly and logically, therefore it is essential that you start at the beginning and work steadily through the text including the derivations, QUESTIONS and ANSWERS and the in-text questions (ITQs). The self-assessment questions (SAQs) should be attempted at the end of the Sections to which they refer. (*Note* You may find differences in arithmetic at times because of the number of decimal points you decide to use.)

You are advised to have the statistics text* handy when reading this Unit as we make direct reference to it in the text.

There are two TV programmes (Heterosis, and Plant and animal breeding) associated with the text both of which deal with the practical problems of breeding plants and animals. Two Radio programmes derive the parameters discussed here and in Unit 12.

Finally, it is difficult to suggest any Sections of Unit 11 that you might reasonably omit without disrupting your appreciation of the subject; the whole Unit will be assessed. However, if you are pressed for time, Sections 11.7 and 11.8 are the best candidates for lighter treatment.

---

* The Open University (1976) S299 STATS *Statistics for Genetics*, The Open University Press. This text is to be studied in parallel with the Units of the Course. We refer to it by its code, *STATS*.

11.0   Introduction to Unit 11

In Units 9 and 10, the operation of natural selection on discontinuous variation was discussed with particular reference to the rate of replacement of alleles and the establishment and maintenance of balanced polymorphisms. In the following three Units we shall consider continuous variation only.

You should of course ask why we need to understand the inheritance of continuously varying characters. It so happens that many such characters are of agronomic, agricultural and evolutionary importance, and it is, therefore, an important class of variation, which we cannot afford to ignore.

For example, consider an important trait such as milk yield in dairy cattle. In a large sample, individual yields may vary over the entire range between, say, 300 and 3 000 litres of milk per cow per year. But let us analyse the components that contribute to this variation. First, environmental differences such as the quality and quantity of food and general animal husbandry will be reflected in the variation. Second, the basic genetic differences among dairy cattle will also contribute to the variation. But, of course, only the genetically determined variation is heritable and, therefore, of use to the animal breeder. Consequently, if the yield of milk from dairy herds is to be improved, or, more generally, if other important characters in domestic species of plants and animals are to be changed for the better, we need to understand the details of their inheritance, and to quantify the relationship between genetic differences and the environment in the generation of continuous variation.

11.1   The historical foundation and basis of continuous variation

Before we discuss the inheritance of continuous variation, it is useful to know some of the background and even controversies that have resulted in an understanding of the patterns of inheritance of continuously varying characters.

continuous variation

In 1900 Mendel's work was rediscovered and its significance understood by de Vries, Correns and von Tschermak-Seysenegg. Although this represented a major breakthrough in elucidating the relationship between parent and offspring, Mendel had deliberately avoided using any character that exhibited continuous variation. Yet continuous variation could not be completely ignored; even Darwin had realized the significance of small cumulative steps in evolutionary change. But before a complete understanding could be achieved, the fusion of two distinct approaches was necessary, namely, the genetic and the statistical or biometrical—the genetic giving us the principles on which any analysis must be based, and the statistical showing us how to describe continuous variation.

Genetics has always been a controversial subject, and in the early days argument raged over the relative importance of the two classes of variation in evolution. Early geneticists such as Bateson and de Vries proposed that continuous variation could not be genetically determined and, therefore, was not heritable; only large discontinuous variations or differences were determined by genes. However, Galton, who was attempting to explain observable differences between humans, noticed that many of the differences were of small, barely perceptible, steps, and that such characters appeared to be inherited. Using biometrical techniques, such as regression and correlation analyses, the biometricians were able to demonstrate statistically the likeness between relatives, even though the segregation and assortment of individual hereditary factors were not separable in the analysis.

The debate over the relative importance of the two classes of variation in evolution resulted in two schools of thought, the Mendelists under Bateson, who believed that all heritable differences that were important in evolution were discontinuous, and the biometricians under Weldon, who equally proposed that heritable variation was basically all continuous. Such was the controversy, that at the beginning of the twentieth century the two schools were deeply entrenched in their respective ideologies.

In 1906 Yule suggested that continuous variation might be produced by many genes, each with a small but equal and cumulative effect on the same character, and in the years that followed, this multiple-factor hypothesis was applied to data from different organisms. It was around this time that R. A. Fisher, in his first paper on this subject, 'The correlation between relatives on the supposition of Mendelian inheritance', showed that the results of the biometricians could be obtained only by

multifactorial inheritance

508

following the fundamental principles of Mendelian genetics. Fisher continued the integration of biometrics with genetics and successfully attempted the first partitioning of continuous variation into the components that the multiple-factor hypothesis led him to expect.

Fisher was deeply concerned that the methods of analysing continuous variation were grossly inadequate and this led him to develop the analysis of variance— arguably the most important analytical tool in genetics and quantitative biology*.

The results obtained by the biometricians in the analysis of continuous variation could be arrived at only by following the fundamental principles of Mendelian genetics, as we mentioned earlier. But how can the intrinsically discontinuous variation that is generated by the segregation of genes be translated into the continuous variation exhibited by many characters?

To answer this question let us first consider a model using a single gene. Consider a hypothetical locus, A, at which there are two alleles, A and a.

In future, we shall refer to this as the A–a gene. For example, suppose that the A–a gene determines flower colour in a variety of tobacco, *Nicotiana tabacum*, where the genotype AA has red flowers and the genotype aa has white flowers.

When the A–a gene segregates in an F_2 generation, we obtain the genotypes AA, Aa and aa in the proportion of $\frac{1}{4} : \frac{1}{2} : \frac{1}{4}$, that is, a relative frequency of $1 : 2 : 1$. If you cannot see how these proportions are derived, remember that at meiosis an individual possessing the gene pair A–a transmits the A allele to half of its gametes and the a allele to the other half. If Aa is mated to another Aa individual to produce an F_2 generation, then the probability that the resulting zygote will be AA is the product of the probabilities that each of the uniting gametes is A. As the probability that either gamete carries the A allele is $\frac{1}{2}$, the probability that the zygote will be AA is given by $P(A) \times P(A) = \frac{1}{2} \times \frac{1}{2} = \frac{1}{4}$, where $P(A)$ is the probability that the gamete carries the A allele.

F_1 genotype		Aa	
F_1 gametes	A		a
Proportion	$\frac{1}{2}$		$\frac{1}{2}$

	AA	Aa	
A $\frac{1}{2}$	$\frac{1}{4}$	$\frac{1}{4}$	F_2 genotypes
a $\frac{1}{2}$	Aa $\frac{1}{4}$	aa $\frac{1}{4}$	

F_2 genotypes in the proportion of $\frac{1}{4}AA : \frac{1}{2}Aa : \frac{1}{4}aa$

If we assume that the A allele is completely dominant to the a allele, then we would obtain a phenotypic ratio of 3 red : 1 white in the F_2, the genotype Aa having the same degree of redness as the AA genotype.

If, on the other hand, the A allele is not dominant to the a allele, and the expression of the heterozygote Aa is midway between the expression of the two homozygotes, then the phenotypic ratio in the F_2 generation would be 1 red : 2 pink : 1 white, corresponding to the genotypes AA, Aa and aa.

So, we can say that where there are no differences in dominance between the two alleles of a gene, the phenotypic value of the heterozygote is dependent on the additive effect of the A and a alleles, that is, the phenotypic contribution of one A allele plus the phenotypic contribution of one a allele.

Let us simplify matters by saying that for every A allele in the genotype, we add a value of half a unit to the phenotypic expression of the character, and for every a allele we subtract half a unit. Thus:

	A $\quad A$	A $\quad a$	a $\quad a$
	$+\frac{1}{2}$ $\ +\frac{1}{2}$	$+\frac{1}{2}$ $\ -\frac{1}{2}$	$-\frac{1}{2}$ $\ -\frac{1}{2}$
phenotype	$+1$	0	-1

---

* The rich and fascinating relationships among early geneticists are discussed in far greater detail in the history text for this Course. (The Open University (1976) S299 HIST *The History and Social Relations of Genetics*, The Open University Press. This text is to be studied in parallel with the Units of the Course. We refer to it by its code *HIST*.)

The value of the heterozygote is dependent on the additive effect of the A and a alleles and is midway between the two homozygotes in expression.

Now, consider two genes, $A–a$ and $B–b$, that segregate independently and determine the expression of the same character.

F₁ genotype			*AaBb*	
F₁ gametes	*AB*	*Ab*	*aB*	*ab*
Proportion	¼	¼	¼	¼

AB $\frac{1}{4}$	*AABB* $\frac{1}{16}$	*AABb* $\frac{1}{16}$	*AaBB* $\frac{1}{16}$	*AaBb* $\frac{1}{16}$	F₂ genotypes
Ab $\frac{1}{4}$	*AABb* $\frac{1}{16}$	*AAbb* $\frac{1}{16}$	*AaBb* $\frac{1}{16}$	*Aabb* $\frac{1}{16}$	
aB $\frac{1}{4}$	*AaBB* $\frac{1}{16}$	*AaBb* $\frac{1}{16}$	*aaBB* $\frac{1}{16}$	*aaBb* $\frac{1}{16}$	
ab $\frac{1}{4}$	*AaBb* $\frac{1}{16}$	*Aabb* $\frac{1}{16}$	*aaBb* $\frac{1}{16}$	*aabb* $\frac{1}{16}$	

We shall again assign a value of half a unit to the phenotypic expression of the character for every A or B allele in the genotype, and subtract half a unit for every a or b allele in the genotype.

Remember that the allele with an increasing effect on the manifestation of the character, which we call an increasing allele, is always designated by a capital letter and the allele of decreasing effect by a small or lower-case letter. In this usage the capital letter does not necessarily imply dominance.

The distribution of phenotypic classes will follow the number of increasing and decreasing alleles present within the different genotypes; this is represented in Table 1.

Table 1   The distribution of phenotypic classes when two genes show independent segregation

phenotypic class	2	1	0	−1	−2
number of genotypes in each class	1	4	6	4	1

Of course, in the way in which we have described this variation, its distribution is still discontinuous.

Now let us go one step further and examine three genes, $A–a$, $B–b$ and $C–c$, that segregate independently and determine the expression of the same character.

QUESTION (a) What are the genotypic ratios in the F₂ generation when three genes, $A–a$, $B–b$ and $C–c$ segregate independently and determine the same character? (b) Assuming that the additive effects of the three genes are equal, that is, $A = B = C = +\frac{1}{2}$, and $a = b = c = -\frac{1}{2}$, and that there are no differences in dominance between the alleles of the genes, construct a table showing the distribution of phenotypic classes.

ANSWER (a) See table at top of opposite page.

(b)

Table 2   The distribution of phenotypic classes when three genes show independent segregation

phenotypic class	3	2	1	0	−1	−2	−3
number of individuals in each class	1	6	15	20	15	6	1

F$_1$ genotype $AaBbCc$

F$_1$ gametes and proportion

	ABC $\frac{1}{8}$	ABc $\frac{1}{8}$	AbC $\frac{1}{8}$	Abc $\frac{1}{8}$	abc $\frac{1}{8}$	abC $\frac{1}{8}$	aBc $\frac{1}{8}$	aBC $\frac{1}{8}$	
ABC $\frac{1}{8}$	$AABBCC$ $\frac{1}{64}$	$AABBCc$ $\frac{1}{64}$	$AABbCC$ $\frac{1}{64}$	$AABbCc$ $\frac{1}{64}$	$AaBbCc$ $\frac{1}{64}$	$AaBbCC$ $\frac{1}{64}$	$AaBBCc$ $\frac{1}{64}$	$AaBBCC$ $\frac{1}{64}$	F$_2$ genotypes
ABc $\frac{1}{8}$	$AABBCc$ $\frac{1}{64}$	$AABBcc$ $\frac{1}{64}$	$AABbCc$ $\frac{1}{64}$	$AABbcc$ $\frac{1}{64}$	$AaBbcc$ $\frac{1}{64}$	$AaBbCc$ $\frac{1}{64}$	$AaBBcc$ $\frac{1}{64}$	$AaBBCc$ $\frac{1}{64}$	
AbC $\frac{1}{8}$	$AABbCC$ $\frac{1}{64}$	$AABbCc$ $\frac{1}{64}$	$AAbbCC$ $\frac{1}{64}$	$AAbbCc$ $\frac{1}{64}$	$Aabbcc$ $\frac{1}{64}$	$AabbCC$ $\frac{1}{64}$	$AaBbCc$ $\frac{1}{64}$	$AaBbCC$ $\frac{1}{64}$	
Abc $\frac{1}{8}$	$AABbCc$ $\frac{1}{64}$	$AABbcc$ $\frac{1}{64}$	$AAbbCc$ $\frac{1}{64}$	$AAbbcc$ $\frac{1}{64}$	$Aabbcc$ $\frac{1}{64}$	$AabbCc$ $\frac{1}{64}$	$AaBbcc$ $\frac{1}{64}$	$AaBbCc$ $\frac{1}{64}$	
abc $\frac{1}{8}$	$AaBbCc$ $\frac{1}{64}$	$AaBbcc$ $\frac{1}{64}$	$AabbCc$ $\frac{1}{64}$	$Aabbcc$ $\frac{1}{64}$	$aabbcc$ $\frac{1}{64}$	$aabbCc$ $\frac{1}{64}$	$aaBbcc$ $\frac{1}{64}$	$aaBbCc$ $\frac{1}{64}$	
abC $\frac{1}{8}$	$AaBbCC$ $\frac{1}{64}$	$AaBbCc$ $\frac{1}{64}$	$AabbCC$ $\frac{1}{64}$	$AabbCc$ $\frac{1}{64}$	$aabbCc$ $\frac{1}{64}$	$aabbCC$ $\frac{1}{64}$	$aaBbCc$ $\frac{1}{64}$	$aaBbCC$ $\frac{1}{64}$	
aBc $\frac{1}{8}$	$AaBBCc$ $\frac{1}{64}$	$AaBBcc$ $\frac{1}{64}$	$AaBbCc$ $\frac{1}{64}$	$AaBbcc$ $\frac{1}{64}$	$aaBbcc$ $\frac{1}{64}$	$aaBbCc$ $\frac{1}{64}$	$aaBBcc$ $\frac{1}{64}$	$aaBBCc$ $\frac{1}{64}$	
aBC $\frac{1}{8}$	$AaBBCC$ $\frac{1}{64}$	$AaBBCc$ $\frac{1}{64}$	$AaBbCC$ $\frac{1}{64}$	$AaBbCc$ $\frac{1}{64}$	$aaBbCc$ $\frac{1}{64}$	$aaBbCC$ $\frac{1}{64}$	$aaBBCc$ $\frac{1}{64}$	$aaBBCC$ $\frac{1}{64}$	

If you experienced difficulty with this QUESTION and ANSWER, remember that the genotype $AABBcc$ would have a net phenotype of 1.

$$
\begin{array}{cccccc}
A & A & B & B & c & c \\
+\tfrac{1}{2} & +\tfrac{1}{2} & +\tfrac{1}{2} & +\tfrac{1}{2} & -\tfrac{1}{2} & -\tfrac{1}{2}
\end{array}
$$
$$\text{net effect} = 1$$

ITQ 1 Using the data from the QUESTION and ANSWER, (a) plot a frequency histogram, and (b) by joining together the top mid-points of the columns, construct a frequency polygon (see *STATS*, Section ST.1.3).

The answers to the ITQs are on p. 550.

From what has been said earlier and from ITQ 1, it should be clear that continuous variation is likely to result from the independent segregation of several genes that affect the same character. It should, however, be equally obvious, that even with 20 or 30 such genes, the various phenotypes would fall unambiguously into precise class intervals. In other words, the distribution would still be essentially discontinuous. How such variation is finally translated into a continuous distribution is determined by the environment, so when the effects of the environment are equal to or greater than the effects of individual genes, then the phenotypic ranges of particular genotypes would tend to overlap, resulting in a smooth, continuous distribution of phenotypes.

The effects of the environment on continuous variation are discussed in greater detail in Section 11.3.1.

Let us summarize the argument developed so far. The essential features of a multi-factorial hypothesis are two-fold:

1 Continuous variation is the result of the independent segregation of many genes that affect the same character. The phenotypic expression of individual genotypes is dependent on the cumulative effects of all the individual genes.

2 When the effects of individual genes are small in relation to non-heritable effects, that is, environmentally induced variation, the discontinuity associated with the segregation of genes becomes undiscernible in the phenotypic distribution.

11.2 The measurement of continuous variation

We have seen that genetic segregation allied with the effects of the environment are a likely explanation of continuous variation. The question that remains is how to quantify or measure such variation.

QUESTION Flowering time in the grass *Agrostis tenuis* is a continuous variate. In Figure 1, the distribution of flowering time in one particular population has been represented as a continuous variate. Can you recall what type of distribution these variates follow?

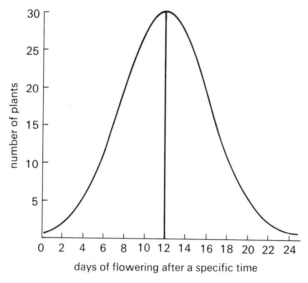

days of flowering after a specific time

Figure 1

ANSWER The distribution of heights follows a bell-shaped, symmetrical curve, which approximates to a normal distribution.

QUESTION Can you recall what properties are used to describe a normal distribution?

ANSWER Two parameters are required to describe a normal distribution, its mean and its variance.

Before continuing with this Unit, you are strongly advised to read STATS, *Section ST.6, where the properties of a normal distribution are discussed in some detail.*

Using the properties of a normal distribution, that is, means, standard deviations and variances, we are able to quantify and, therefore, to understand the patterns of inheritance of continuously varying characters.

11.3 Components of generation means

In the measurement of discontinuous variation, individuals are assignable to one or other of a number of well-defined and distinct classes. For example, if we examine briefly three of the characters used by Mendel in his classic experiments with the pea *Pisum sativum*, any plant falls readily into one group or the other.

1 Round seeds or wrinkled seeds.

2 Long stems or short stems.

3 White testa or grey–brown testa.

Let us take a closer look at the character, stem length. Mendel assigned plants that were between 6 and 7 feet in length to the tall class and those between $\frac{3}{4}$ and $1\frac{1}{2}$ feet to the short class. Clearly, the precise metrical relationship between the two groups of phenotypes is irrelevant; what is important is to be able to distinguish between tall and short stemmed plants unambiguously.

With continuous variation, however, each observation is unique, the total range of phenotypes having a smooth distribution without regular discontinuities. Each measurement, therefore, gives the precise level of expression of a character and the precise metrical relationship between phenotypes is important.

In this Section we shall consider a way in which the metrical relationship between phenotypes can be described in terms of biometrical quantities, means or averages.

11.3.1 Environmental effects and the need for randomization

Let us suppose that we wish to evaluate a new, pure-breeding (or true-breeding), variety of barley. We plant, say, 1 000 seeds, and some months later harvest the crop. If the seeds from individual plants are weighed and a frequency histogram of the data is constructed, a continuous distribution of seed weights would be obtained even though the individual plants themselves are genetically identical.

environmental effects

Such a distribution of phenotypes can be attributed only to differences in environmental conditions to which the individual plants have been subjected.

Even though the experiment may have been performed in an environment that has been made as uniform as practicable, local, uncontrolled or uncontrollable, differences of many factors, for example, the availability of nutrients in the soil, humidity, the intensity of light and competition, will lead to variation among individuals.

An important point to remember at this stage, is that when we measure the phenotype, for instance, grams of barley per plant, it is only the mean phenotype corresponding to a particular set of conditions. If any of these are changed, for instance, by increasing the application of nitrogen, the mean phenotype would alter accordingly. Unless, however, all the individuals are grown in identical conditions within any one environment, which is of course practically impossible, the distribution of phenotypes will be continuous even though the mean itself will change.

Many plant and animal breeders are interested in comparing the performance of new varieties with trusted and well-tried older ones. If, however, so much variation can occur within a single genotype, there is a very real problem of deciding whether differences between genotypes or varieties are the result of genetic differences or merely due to the environment.

For example, two true-breeding varieties of barley have been grown according to the experimental design shown in Figure 2.

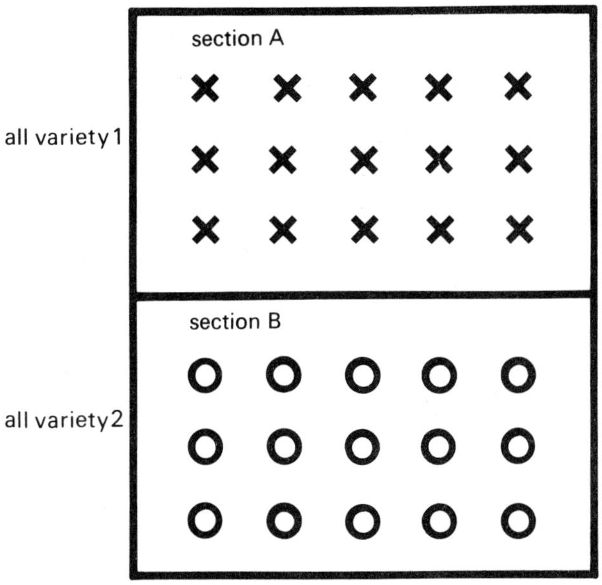

experimental design

Figure 2

When seed from the two varieties is harvested, variety 1 has a mean yield of 25 grams per plant, whereas variety 2 has a mean yield of 55 grams per plant.

QUESTION Do you think it is reasonable to conclude that 2 is a more productive variety than 1 because of genetic differences?

ANSWER This may well be, but such a conclusion cannot be drawn from the information given above. For instance, it is equally likely that section A of the experimental field is entirely different from section B. If this were so, then the two values would be estimates of the mean phenotypes in two different and, therefore, not comparable environments.

QUESTION Can you suggest an experimental design that would ensure that the two varieties were dispersed in the experimental plot in such a way that, as far as possible, each variety would be subjected to the same range of environmental conditions?

ANSWER

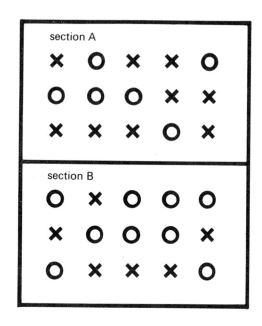

Figure 3

In Figure 3 we have presented an experimental design that would ensure that each variety would be subjected to the same range of environmental conditions. Individuals of both varieties have been distributed at random within the experimental area as a whole, thus ensuring that individuals of any one variety will not be subjected to a different and exclusive range of environmental conditions.

Randomization of experimental material is not difficult, and we do not need to discuss it any further. *You are advised, however, to read Section ST.9.2 of* STATS *at this stage.*

The important point to remember is that whenever we compare different genotypes or different varieties, we shall have assumed that many individuals of each variety have been grown together in a randomized experimental design and that their mean phenotypes are directly comparable. Similarly, the genetic differences that we attempt to quantify will be in terms of the average effects of the genes over the range of environments in which the experimental material has been raised.

11.3.2 Additive effects

We shall begin by considering a simple case and then extend the general principles into more complicated, and more realistic, situations. Take one gene, say, the $A-a$ gene that determines the height of straw in a variety of barley. It is possible to derive three genotypes from the $A-a$ gene, namely,

$$AA \qquad Aa \qquad aa$$

In order to measure the relationship between the three genotypes, we 'build a model' that expresses gene differences in terms of the additive and dominance properties of the alleles of the gene.

If we assume no difference in dominance between the two alleles of the gene, then the genotype AA can be considered to be Aa plus the additive effect of one A allele,

514

and the genotype aa can be considered to be Aa minus the additive effect of one A allele. Representing this diagrammatically, we have:

$$AA \qquad\qquad\qquad Aa \qquad\qquad\qquad aa$$
$$\longleftarrow \quad +A \quad \longrightarrow \longleftarrow \quad -A \quad \longrightarrow$$

Now, we shall denote the additive effect of the A allele by d. As we are referring to the A–a gene, we can identify d by adding the subscript a, that is d_a. Similarly, if we were referring to the B–b, C–c or D–d genes, their additive effects would be given by d_b, d_c and d_d, respectively.

$$AA \qquad\qquad\qquad Aa \qquad\qquad\qquad aa$$
$$\longleftarrow \quad +d_a \quad \longrightarrow \longleftarrow \quad -d_a \quad \longrightarrow$$

As Aa has the additive effect of one A allele and one a allele, then Aa is midway in expression between the two homozygotes and is termed the mid-parent value, denoted by the letter, m.

The *mid-parent value*, m, is the natural mid-point from which to measure the deviation of the two homozygotes, the deviation being equal to the additive effect (d_a) of the gene.

mid-parent value, m

$$AA \qquad\qquad\qquad m \qquad\qquad\qquad aa$$
$$\longleftarrow \quad +d_a \quad \longrightarrow \longleftarrow \quad -d_a \quad \longrightarrow$$

It also follows that the relationship AA, aa and m, can be expressed in terms of:

$$AA = m + d_a$$
$$aa = m - d_a$$

How can we calculate m?

Two true-breeding lines of barley differ genetically in being homozygous for different alleles of the A–a gene that determines straw length. 40 individuals of each genotype are grown in a randomized experimental design and the mean straw lengths are found to be:

$$AA = 150 \text{ cm}$$
$$aa = 50 \text{ cm}$$

Now, the A–a gene is responsible for the difference between the measured values of the two homozygotes. As AA measures 150 cm and aa measures 50 cm, the difference between the 2 is equal to 100 cm. We can say, therefore, that 2 A alleles give a length of 100 cm and that 1 A allele is responsible for increasing the length by 50 cm. Another way of looking at this is to say that 2 a alleles reduce the length by 100 cm and that 1 a allele is responsible for reducing the length by 50 cm.

$$\begin{array}{ccc} AA & m & aa \\ 150 & 100 & 50 \end{array}$$
$$\longleftarrow \quad +d_a = 50 \quad \longrightarrow \longleftarrow \quad -d_a = -50 \quad \longrightarrow$$

By summing the additive effects of 1 A allele and 1 a allele, that is, $+50$ cm and -50 cm, the contribution of the A–a gene is effectively eliminated.

Of course, a far easier way of calculating m is to add the values of both homozygotes and then divide by two!

$$m = \frac{AA + aa}{2} = \frac{150 + 50}{2} = 100 \text{ cm}$$

QUESTION Using the same data, calculate d_a.

ANSWER As AA is equal to $m + d_a$, it follows that d_a is equal to $AA - m$. That is, d_a is equal to $150 - 100 = 50$ cm. Therefore, $d_a = 50$ cm.

$$\begin{array}{ccc} AA & m & aa \\ 150 & 100 & 50 \end{array}$$
$$\longleftarrow \quad +d_a = 50 \quad \longrightarrow \longleftarrow \quad -d_a = -50 \quad \longrightarrow$$

QUESTION Two true-breeding lines of oats differ genetically in being homozygous for different alleles of the same gene. Assuming normal experimental practice, calculate m and d_b, where BB yields 80 g and bb 30 g of seed.

ANSWER
$$m = \frac{BB + bb}{2} = \frac{80 + 30}{2} = 55 \text{ g}$$

It is very important to remember that when we are quantifying the metrical relationship between the measurable properties of different genotypes, we can measure only the *differences* between such lines. It is more than likely that the majority of plants and animals have many genes that affect the same character but several of these genes will be present in the same state and will not, therefore, contribute to the measurable differences between the animals and plants. In other words, we are trying to assess the effects of a difference in a single gene in the two genotypes against the background of the remaining genes, which, because they are common to both genotypes, do not cause measurable differences between them.

The genes that are not different, however, will affect the expression of the character in *both* genotypes to the *same* extent. For example,

$$
\begin{array}{ll}
\text{genotype 1} & \boxed{BBCCDD \;\; AA} \\
\text{genotype 2} & \boxed{BBCCDD \;\; aa}
\end{array}
$$

Any differences between the measurable properties of the two genotypes must be entirely due to the A–a gene. Clearly, the BB, CC and DD genes will contribute to the expression of the character in both genotypes to the same extent.

We now need to quantify the metrical relationship between AA and aa within the framework of the other *common* genes.

We shall assume that each AA, BB, CC and DD gene contributes an increment of 4 cm to the measurable properties of the genotypes, whereas aa produces a reduction of 4 cm. That AA contributes $+4$ and aa -4 follows, as by definition the relationship between AA, m and aa is given by:

$$
\underset{\longleftarrow \quad +d_a \quad \longrightarrow}{AA} \quad \underset{\longleftarrow \quad -d_a \quad \longrightarrow}{m} \quad aa
$$

$$
\begin{array}{l}
\text{genotype 1} \quad \boxed{
\begin{array}{cccccc|cc}
B & B & C & C & D & D & A & A \\
+2 & +2 & +2 & +2 & +2 & +2 & +2 & +2 = 16 \text{ cm}
\end{array}} \\[2em]
\text{genotype 2} \quad \boxed{
\begin{array}{cccccc|cc}
B & B & C & C & D & D & a & a \\
+2 & +2 & +2 & +2 & +2 & +2 & -2 & -2 = 8 \text{ cm}
\end{array}}
\end{array}
$$

In the same way in which we obtained the mid-parent value of the genotypes AA and aa, that is, by summing the two and then dividing by two, we have here:

$$
\frac{\text{genotype 1} + \text{genotype 2}}{2} = \text{the mid-parent value, } m
$$

QUESTION What is the numerical value of m?

ANSWER
$$
m = \frac{16 + 8}{2} = 12 \text{ cm.}
$$

Let us now look at what we have done genetically:

$$
\begin{array}{l}
\text{genotype 1} \quad \boxed{
\begin{array}{cccccccc}
B & B & C & C & D & D & A & A \\
+2 & +2 & +2 & +2 & +2 & +2 & +2 & +2
\end{array}} = 16 \text{ cm} \\[2em]
\text{genotype 2} \quad \boxed{
\begin{array}{cccccccc}
B & B & C & C & D & D & a & a \\
+2 & +2 & +2 & +2 & +2 & +2 & -2 & -2
\end{array}} = 8 \text{ cm} \\[2em]
\dfrac{G_1 + G_2}{2} \quad \boxed{
\begin{array}{cccccccc}
B & B & C & C & D & D & A & a \\
+2 & +2 & +2 & +2 & +2 & +2 & +2 & -2
\end{array}} = 12 \text{ cm}
\end{array}
$$

When we add the values of the two genotypes and divide by two, we have effectively changed *only* the A–a gene. As the additive effect of 1 A allele is equal to $+2$ cm and that of 1 a allele is -2 cm, the expression of the A–a gene has been effectively cancelled out.

Consequently, the mid-parent value, m, is the effect of all genes shared commonly by both genotypes, and it is, therefore, the natural mid-point from which to measure the deviation of the homozygotes and thus quantify their metrical relationship. From Section 11.3.1 it should be clear that m also reflects the common environment in which the genotypes have been raised.

Consider an example that we have already used in which two lines of barley were found to have straw lengths of

$$AA = 150 \text{ cm}$$
$$aa = 50 \text{ cm}$$
$$m = \frac{150 + 50}{2} = 100 \text{ cm}$$

Thus, we have:

$$
\begin{array}{ccc}
AA & m & aa \\
\xleftarrow{} +d_a = 50 \xrightarrow{} \xleftarrow{} -d_a = -50 \xrightarrow{} & & \\
\end{array}
$$

$$150 \qquad 100 \qquad 50 \qquad 0$$

where m reflects the value of all the genes, other than those at the A–a locus, that are shared in common by the two genotypes, plus the common environmental effects.

Consider another numerical example. The weights of seeds derived from two true-breeding lines of barley that differ genetically in being homozygous for different alleles of the C–c gene are

$$CC = 58 \text{ g}$$
$$cc = 26 \text{ g}$$

QUESTION Calculate the mid-parent value, m.

ANSWER
$$m = \frac{58 + 26}{2} = 42 \text{ g}$$

QUESTION Calculate the additive effect of the C–c gene, d_c.

ANSWER
$$d_c = 58 - 42 = 16 \text{ g}$$

$$
\begin{array}{ccc}
CC & m & cc \\
58 & 42 & 26 \\
\xleftarrow{} + d_c = 16 \xrightarrow{} \xleftarrow{} -d_c = -16 \xrightarrow{} & & \\
\end{array}
$$

The important point to notice here is that the value of d is *always positive*. The genotype cc is related to m by $-d_c$, but d_c itself is equal to 16 g. In other words, it is impossible to have a negative additive effect.

> ITQ 2 Suppose that the weights of seeds in two lines of barley are determined by the K–k locus. The two lines differ genetically in being homozygous for different alleles of the K–k gene, and the mean measurements given below were obtained from a suitably designed experiment.
>
> $$KK = 48 \text{ g}$$
> $$kk = 26 \text{ g}$$
>
> Calculate the mid-parent value and the additive effect of the K–k gene, d_k.

11.3.3 Dominance effects

So far we have avoided the complication of dominance effects by considering the additive effects of the genes in isolation. Now, in a cross between two lines that are homozygous for different alleles of the A–a gene that determines the length of straw in a variety of oats, the F_1 heterozygote is actually found to be different from the mid-parent value, m:

dominance

$$AA = 100 \text{ cm}$$
$$aa = 50 \text{ cm}$$
$$\text{the measured value of the } F_1 \text{ heterozygote} = 100 \text{ cm}$$

Now, m is again given by $\dfrac{100 + 50}{2} = 75$ cm, and this is the expected value of one A allele and one a allele, considering only the additive effects. But the measured value of Aa is 100 cm. What does this tell us about the dominance relationship of the two alleles of the A–a gene? As the heterozygote has the same value as the AA genotype, this clearly indicates that the A allele is showing *complete* dominance over the a allele.

Let us now suppose that the heterozygote has a measured value of 80 cm. By measuring the deviation of the measured value of the heterozygote from the mid-parent value, the dominance effect of the gene can be quantified.

AA		Aa	m		aa
100		80	75		50

$$\longleftarrow +5 \longrightarrow$$

$$\longleftarrow \quad +d_a = 25 \quad \longrightarrow \longleftarrow \quad -d_a = -25 \quad \longrightarrow$$

Therefore, the A allele is *incompletely* dominant to the a allele in this example.

The level of dominance is denoted by the symbol h. As we are concerned with the A–a gene, we identify h with the subscript a, that is, h_a. Similarly, the dominance levels of the B–b, C–c and D–d genes would be referred to as h_b, h_c and h_d, respectively.

AA		Aa	m		aa
100		80	75		50

$$\leftarrow h_a = +5 \rightarrow$$

$$\longleftarrow \quad +d_a = 25 \quad \longrightarrow \longleftarrow \quad -d_a = -25 \quad \longrightarrow$$

Here is another example illustrating these new points. The weight of seeds in a variety of barley is determined by the A–a gene and the mean weights of the two homozygous lines and their F_1 heterozygote are

$$AA = 55 \text{ g}$$
$$aa = 20 \text{ g}$$
$$Aa = 48 \text{ g}$$

QUESTION Calculate the mid-parent value, m.

ANSWER
$$m = \frac{55 + 20}{2} = 37.5 \text{ g}$$

QUESTION Calculate d_a.

ANSWER
$$d_a = 55 - 37.5 = 17.5$$

QUESTION Calculate h_a.

ANSWER
$$h_a = 48 - 37.5 = +10.5$$

So we have:

AA	Aa	m		aa
55	48	37.5		20

$$\longleftarrow h_a = +10.5 \longrightarrow$$

$$\longleftarrow \quad +d_a = 17.5 \quad \longrightarrow \longleftarrow \quad -d_a = -17.5 \quad \longrightarrow$$

11.3.4 Incomplete dominance

In the last example, the A allele showed dominance over the a allele, but the dominance was not complete; otherwise the value of the heterozygote, Aa, would have been equal to that of the AA parent.

A situation in which dominance is not complete but is in a specific direction is termed *incomplete dominance*. Let us give a numerical example using purely illustrative data.

incomplete dominance

$$AA = 50$$
$$Aa = 35$$
$$aa = 10$$

So, we have:

AA		Aa	m		aa
50		35	30		10

$$\leftarrow h_a = +5 \rightarrow$$

$$\longleftarrow \quad +d_a = 20 \quad \longrightarrow \longleftarrow \quad -d_a = -20 \quad \longrightarrow$$

Here, the dominance effect, h_a is equal to $+5$; that is, the deviation of the F_1 heterozygote from the mid-parent value is 5 units. As the dominance is in the increasing direction (remember it is our convention that the increasing allele is always represented by a capital letter), the value of h_a takes a positive sign.

Using the symbols for additive and dominance effects, we can say that when A is incompletely dominant to a, h_a is smaller than $+d_a$.

On the other hand, look at this data:

$$AA = 50$$
$$Aa = 20$$
$$aa = 10$$

AA		m	Aa	aa
50		30	20	10

$\longleftarrow h_a = -10 \longrightarrow$

$\longleftarrow +d_a = 20 \longrightarrow \longleftarrow -d_a = -20 \longrightarrow$

The dominance effect, h_a, is equal to 10 units. However, as the direction of dominance is towards the decreasing allele, the value of h_a takes a negative sign.

QUESTION Using the symbols for additive and dominance effects, show the relationships between h_a and d_a in the following situations:

(a) Neither allele of the gene is dominant.

(b) The A allele is completely dominant to the a allele.

(c) The a allele is completely dominant to the A allele.

ANSWER

(a)

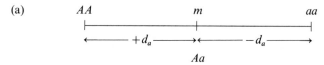

The heterozygote will have the same value as m and therefore $h_a = 0$.

(b)

The heterozygote will have the same value as the AA parent and therefore h_a is positive and equal to d_a.

(c)

The heterozygote will have the same value as the aa parent and therefore h_a is negative and equal to d_a.

In the same way as we defined $\quad AA$ as being equal to $m + d_a$

and $\qquad\qquad\qquad\qquad\qquad aa$ as being equal to $m - d_a$

we can define the heterozygote Aa as being equal to $m + h_a$.

You should recall, specifically from Units 9 and 10, that dominance is the property of characters and *not* of genes. It is important to remember that when we use the words 'the A allele is showing dominance to the a allele', we are using a convenient shorthand for 'the phenotype determined by the A allele is dominant to the phenotype determined by the a allele'.

ITQ 3 Two lines of barley, which differ genetically in being homozygous for different alleles of the same gene, and their F_1 heterozygote are grown in appropriate experimental conditions. The mean weights of the seed produced by the three genotypes are

$$\text{genotype } BB = 84 \text{ g}$$
$$\text{genotype } Bb = 60 \text{ g}$$
$$\text{genotype } bb = 16 \text{ g}$$

Calculate m, d and h.

We can summarize the dominance relationships of a pair of alleles in terms of the ratio of h to d. This is called the *degree of dominance*. The degree of dominance can be expressed in terms of $\dfrac{h}{d}$ or, in ITQ 3 specifically, $\dfrac{h_b}{d_b}$.

degree of dominance

QUESTION   In ITQ 3, d_b was found to be 34, and h_b was found to be $+10$. Calculate the degree of dominance.

ANSWER   The degree of dominance $= \dfrac{h_b}{d_b} = \dfrac{+10}{34} = +0.29.$

The plus sign indicates that the parent with the increasing gene is dominant to the parent with the decreasing genes.

11.3.5   *m*, *d* and *h* as components of generation means

Parental types of generations are referred to as P_1 and P_2; in all cases, the parent with the greater value is termed the P_1 and the parent with the smaller value is termed the P_2. Thus, if the genotype AA were 200 cm high, and aa 100 cm high, AA would become the P_1 parent and aa the P_2 parent.

In Section 11.3.1, it was specified that many individuals of each genotype would be grown in a randomized experimental design, and that any measurements that we discussed would be the means of several individuals. In other words m, d and h are average effects over the range of environmental conditions in which the material has been grown.

Assuming a difference in one gene, then, we can say that the mean of the P_1 generation, which is given the symbol $\overline{P}_1$ [spoken, P_1 bar], is equal to $m + d$. That is:

$$\overline{P}_1 = m + d$$

Similarly,

$$\overline{P}_2 = m - d$$

and

$$\overline{F}_1 = m + h$$

We refer to m, d and h as *components of the generation means*: m being the mid-parent value, d being the additive component and h being the dominance component.

components of generation means
additive component, dominance component

In fact, three further generations are of interest to us, although the reason for this will not be apparent until later. They are:

1   The F_2 generation, that is, $F_1 \times F_1$.
2   The backcross generation $F_1 \times P_1$, which is termed the B_1 generation.
3   The backcross generation $F_1 \times P_2$, which is termed the B_2 generation.

backcross generations

As these three generations undergo segregation, it is necessary to derive the proportion of the different genotypes and their relative contributions to their respective generation means.

Let us begin by looking at the F_2 generation. When the A–a gene segregates in an F_2 generation, we obtain the genotypes AA, Aa and aa in the proportion of $\frac{1}{4} : \frac{1}{2} : \frac{1}{4}$.

F_1 gametes	A	a	
Proportion	$\frac{1}{2}$	$\frac{1}{2}$	
A $\frac{1}{2}$	AA $\frac{1}{4}$	Aa $\frac{1}{4}$	F_2 genotypes
a $\frac{1}{2}$	Aa $\frac{1}{4}$	aa $\frac{1}{4}$	

And the F_2 generation is composed of $\frac{1}{4}AA : \frac{1}{2}Aa : \frac{1}{4}aa$.

(*Note*   If you cannot recall how these frequencies are arrived at, return to Section 11.1.)

In terms of the parameters m, d and h, we can say that the genotype AA has a mean value of $m + d$. But only one quarter of the F_2 generation are AA genotypes.

Consequently, AA contributes $\frac{1}{4}(m + d)$ to the F_2 generation mean. Similarly,

$$Aa \text{ contributes } \tfrac{1}{2}(m + h)$$
$$aa \text{ contributes } \tfrac{1}{4}(m - d)$$

Summing these proportions we obtain the mean of the F_2 generation as $m + \frac{1}{2}h$.

That is, $$\overline{F}_2 = m + \tfrac{1}{2}h$$

QUESTION Using the same procedure as we used for deriving the mean of the F_2 generation, calculate the mean of (a) the B_1 and (b) the B_2 generations. (Hint: $B_1 = Aa \times AA$, and $B_2 = Aa \times aa$.)

ANSWER

(a)

B_1 gametes	A	a
Proportion	$\frac{1}{2}$	$\frac{1}{2}$

	A	a
A $\frac{1}{2}$	AA $\frac{1}{4}$	Aa $\frac{1}{4}$
A $\frac{1}{2}$	AA $\frac{1}{4}$	Aa $\frac{1}{4}$

And the B_1 generation is composed of $\frac{1}{2}Aa : \frac{1}{2}AA$.

In terms of m, d and h,

$$AA \text{ contributes } \tfrac{1}{2}(m + d)$$
$$Aa \text{ contributes } \tfrac{1}{2}(m + h)$$

Summing these proportions, we obtain the mean of the B_1 generation as $m + \frac{1}{2}d + \frac{1}{2}h$.

That is, $$\overline{B}_1 = m + \tfrac{1}{2}d + \tfrac{1}{2}h$$

(b)

B_2 gametes	A	a
Proportion	$\frac{1}{2}$	$\frac{1}{2}$

	A	a
a $\frac{1}{2}$	Aa $\frac{1}{4}$	aa $\frac{1}{4}$
a $\frac{1}{2}$	Aa $\frac{1}{4}$	aa $\frac{1}{4}$

And the B_2 generation is composed of $\frac{1}{2}Aa : \frac{1}{2}aa$.

In terms of m, d and h

$$Aa \text{ contributes } \tfrac{1}{2}(m + h)$$
$$aa \text{ contributes } \tfrac{1}{2}(m - d)$$

Summing these proportions, we obtain the mean of the B_2 generation as $m + \frac{1}{2}h - \frac{1}{2}d$.

That is, $$\overline{B}_2 = m - \tfrac{1}{2}d + \tfrac{1}{2}h$$

Summary

In terms of a single gene difference, the components of the generation means are given by:

$$\overline{P}_1 = m + d$$
$$\overline{P}_2 = m - d$$
$$\overline{F}_1 = m + h$$
$$\overline{F}_2 = m + \tfrac{1}{2}h$$
$$\overline{B}_1 = m + \tfrac{1}{2}d + \tfrac{1}{2}h$$
$$\overline{B}_2 = m - \tfrac{1}{2}d + \tfrac{1}{2}h$$

You should now attempt SAQ 1 (p. 549).

11.4 Generation means—differences in many genes

So far, we have considered the situation in which a difference in only one gene determines the level of expression of a character. We have seen that by using lines that differ genetically in being homozygous for different alleles of a single gene and successive generations derived from a cross between the lines, the parameters m, d and h may be determined, and that this accounts for all the differences in generation means.

It should be obvious, however, that continuous variation is unlikely to be determined by the segregation of single genes; it is necessary, therefore, to consider the situation in which more than one gene determines the expression of a character.

Just as for a difference in a single gene, we need to quantify the metrical relationship between the various genotypes, and this can again be considered in terms of the mid-parent value, and the additive and dominance effects of the genes involved.

11.4.1 Additive effects

When more than one gene determines a character, each gene has an independent additive effect, that is, a d effect, on the expression of the character.

If we take any two true-breeding lines, for example,

$$AA\,BB\,CC \quad \text{and} \quad aa\,bb\,cc$$

in which the A, B and C alleles are increasing the manifestation of the character, and the a, b and c alleles are decreasing it, then the additive effects of the individual genes can be represented as:

$$
\begin{array}{ccccccc}
AA & BB & CC & & aa & bb & cc \\
| & | & | & & | & | & | \\
| & | & | & & | & | & | \\
+d_a & +d_b & +d_c & & -d_a & -d_b & -d_c
\end{array}
$$

We can say that the additive effect of the genes is equal to the sum total of all the individual additive effects of each gene. This is denoted by $\sum$, which means 'sum of'.

Thus,

$$
\begin{array}{ccc}
AA & BB & CC \\
| & | & | \\
| & | & | \\
+d_a & +d_b & +d_c = \sum(+d_a + d_b + d_c) = \sum(d_+)
\end{array}
$$

d_+ is used because they are all increasing alleles.

Similarly,

$$
\begin{array}{ccc}
aa & bb & cc \\
| & | & | \\
| & | & | \\
-d_a & -d_b & -d_c = \sum(-d_a - d_b - d_c) = \sum(d_-)
\end{array}
$$

d_- is used because they are all decreasing alleles.

When all the alleles of like effect are in the one strain, as in the example above, we have *complete association*.

complete association

You may have already realized how we can treat genotypes in which all the increasing or decreasing alleles are *not* together in the one strain.

Just to make sure that you are correct: consider a genotype $AA\,BB\,cc$, in which there are two increasing alleles and one decreasing allele.

$$
\begin{array}{ccc}
AA & BB & cc \\
| & | & | \\
| & | & | \\
+d_a & +d_b & -d_c
\end{array}
$$

$+d_a$ and $+d_b$ have increasing additive effects on the expression of the character; $-d_c$ has a decreasing additive effect. We can say, therefore, that the phenotypic expression of a true-breeding line is a balance between the increasing and decreasing alleles.

$$
\begin{array}{ccc}
AA & BB & cc \\
| & | & | \\
| & | & | \\
+d_a & +d_b & -d_c = \sum(d_+) - \sum(d_-)
\end{array}
$$

where $\sum(d_+)$ is the summed additive effects of all the increasing alleles, and $\sum(d_-)$ is the summed effects of all the decreasing alleles.

QUESTION What are the values of $\sum(d_+)$ and $\sum(d_-)$ in the following examples? For the sake of simplicity, we shall assume equal additive effects, so that $+d_a = +d_b = +d_c = +d_d = +d_e = +d_f = 1$.

AA BB CC DD EE FF

AA bb cc DD EE FF

AA BB cc dd ee FF

ANSWER

	$\sum(d_+)$	$\sum(d_-)$
AA BB CC DD EE FF	6	0
AA bb cc DD EE FF	4	2
AA BB cc dd ee FF	3	3

If you had difficulty with this QUESTION and ANSWER, remember that

$$
\begin{array}{ll}
AA = +d_a & aa = -d_a \\
BB = +d_b & bb = -d_b \\
CC = +d_c & cc = -d_c \\
DD = +d_d & dd = -d_d \\
EE = +d_e & ee = -d_e \\
FF = +d_f & ff = -d_f
\end{array}
$$

We have seen that the phenotypic expression of a character in a true-breeding line is a balance between the additive effects of the increasing and decreasing alleles. That is, $\sum(d_+) - \sum(d_-)$. This balance effect can be denoted by $[d]$, where $[d] = \sum(d_+) - \sum(d_-)$.

QUESTION Calculate $[d]$ for the examples in the last QUESTION and ANSWER.

ANSWER

	$\sum(d_+)$	$\sum(d_-)$	$[d]$
AA BB CC DD EE FF	6	0	6
AA bb cc DD EE FF	4	2	2
AA BB cc dd ee FF	3	3	0

If we now consider two strains that are homozygous for different alleles of the same genes, the two strains can be defined as:

$$
\overline{P}_1 = m + [\sum(d_{+1}) - \sum(d_{-1})]
$$
$$
\overline{P}_2 = m + [\sum(d_{+2}) - \sum(d_{-2})]
$$

which has algebraic equivalence to

$$
\overline{P}_2 = m - [\sum(d_{-2}) - \sum(d_{+2})]
$$

As we are concerned with the differences between pairs of lines that are homozygous for different alleles of the same genes, it follows that

$$
\sum(d_{+1}) = \sum(d_{-2})
$$

and

$$
\sum(d_{-1}) = \sum(d_{+2})
$$

We can, therefore, rewrite the above definitions as

$$\overline{P}_1 = m + [\sum (d_{+1}) - \sum (d_{-1})]$$
$$\overline{P}_2 = m - [\sum (d_{+1}) - \sum (d_{-1})]$$

As $\sum (d_+) - \sum (d_-) = [d]$, we have

$$\overline{P}_1 = m + [d]$$
$$\overline{P}_2 = m - [d]$$

and the difference between the two strains is $2[d]$.

Let us take a numerical example. In two true-breeding parental lines of barley the straw length provides the following values. We shall again assume that these measurements have been made in appropriate experimental designs.

$$\overline{P}_1 = 800 \text{ mm}$$
$$\overline{P}_2 = 400 \text{ mm}$$

QUESTION Calculate the mid-parent value, m.

ANSWER m is calculated in exactly the same way as for a single gene difference.

Thus:

$$m = \frac{\overline{P}_1 + \overline{P}_2}{2} = \frac{800 + 400}{2} = 600$$

QUESTION Calculate $[d]$.

ANSWER $[d]$ is calculated in exactly the same way as for d. Thus

$$[d] = \overline{P}_1 - m = 800 - 600 = 200$$

Therefore, the additive genetic effect is equal to 200

ITQ 4 Two pure-breeding lines of wheat have mean yields of seed of 80 g and 120 g, respectively. Calculate m and $[d]$.

11.4.2 Dominance effects

As we saw in Section 11.3.3, the value of h may take a positive or a negative sign according to which allele of the gene is dominant. When many genes determine a character, the observed dominance effects are again a balance between the h effects in an increasing direction and the h effects in a decreasing direction.

Representing the balance effect by $[h]$, we have $[h] = \sum (h_+) - \sum (h_-)$, where $\sum (h_+)$ is the sum of the dominance effects of dominant increasing alleles, and $\sum (h_-)$ is the sum of the dominance effects of dominant decreasing alleles.

Because the values of the independent h items take a positive or negative sign, $[h]$ is also equal to $\sum (h)$. To illustrate this point, consider the heterozygote $AaBbCcDd$, where $h_a = h_b = h_c = h_d = +1$

$$[h] = \sum (h_+) - \sum (h_-) = 4$$

$$\sum (h) = h_a + h_b + h_c + h_d = 1 + 1 + 1 + 1 = 4$$

Now, consider the heterozygote $AaBbCcDd$, where $h_a = h_b = h_c = +1$, and $h_d = -1$.

$$[h] = \sum (h_+) - \sum (h_-) = 3 - 1 = 2$$

$$\sum (h) = h_a + h_b + h_c + h_d = 1 + 1 + 1 + (-1) = 2$$

Thus, $\sum (h)$ and $[h]$ are exactly the same. The important thing is that $\sum (h)$ is *not* the total dominance effect of the genes, but the balance between those h effects in an increasing direction and those h effects in a decreasing direction. If, however, all the independent h items have the same sign—that is, dominance is unidirectional— then $\sum (h)$ is the total dominance effect of the genes.

Consider the genotype $AaBbCcDd$, where $h_a = h_b = h_c = h_d = +1$

$$
\begin{array}{cccc}
Aa & Bb & Cc & Dd \\
| & | & | & | \\
| & | & | & | \\
h_a = +1 & h_b = +1 & h_c = +1 & h_d = +1 \\
\end{array}
$$
$$[h] = 4$$

$\sum (h) = 4$, and in this example $\sum (h)$ *is* the total dominance effect of the genes.

Now consider the genotype $AaBbCcDd$, where $h_a = h_b = h_c = +1$, and $h_d = -1$.

$$
\begin{array}{cccc}
Aa & Bb & Cc & Dd \\
| & | & | & | \\
| & | & | & | \\
h_a = +1 & h_b = +1 & h_c = +1 & h_d = -1 \\
\end{array}
$$
$$[h] = 2$$

$\sum (h) = 2$, and in this example $\sum (h)$ *is not* the total dominance effect of the genes.

QUESTION The means of the P_1 and P_2 generations have been defined in terms of m and $[d]$:
$$\bar{P}_1 = m + [d]$$
$$\bar{P}_2 = m - [d]$$

Can you see how the mean of the F_1 generation may be defined?

ANSWER The mean of the F_1 generation may be defined in terms of m and $[h]$:
$$\bar{F}_1 = m + [h]$$

The value of $[h]$ can take a positive or a negative sign, depending on the balance between the h effects in an increasing direction and the h effects in a decreasing direction. Thus, the deviation of $\bar{F}_1$ from the mid-parent value may be positive or negative.

For illustrative purposes only, let us assume that
$$\bar{P}_1 = 50$$
$$\bar{P}_2 = 20$$
$$\bar{F}_1 = 40$$
$$m = \frac{50 + 20}{2} = 35$$
$$[d] = 50 - 35 = 15$$
$$[h] = 40 - 35 = +5$$

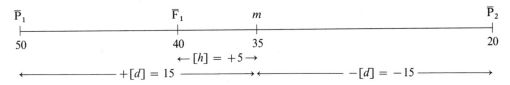

Using the same figures for $\bar{P}_1$ and $\bar{P}_2$, we shall now let $\bar{F}_1 = 30$
$$m = 35$$
$$[d] = 15$$
$$[h] = 30 - 35 = -5$$

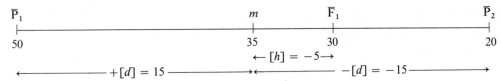

ITQ 5 The values of the mean yields of grain in two true-breeding lines of oats and their F_1 heterozygote are given below. Calculate the parameters m, $[d]$ and $[h]$.

$$\bar{P}_1 = 15.0 \text{ g}$$
$$\bar{P}_2 = 10.0 \text{ g}$$
$$\bar{F}_1 = 11.0 \text{ g}$$

11.4.3 m, $[d]$ and $[h]$ as components of generation means

In Section 11.3.5, we defined successive generations in terms of the parameters m, d and h, and we called these 'components of the generation means' because the mean value of any generation could be defined in terms of m, d and h.

In the same way, when more than one gene difference determines a character, we can define the generation means in terms of the parameters m, $[d]$ and $[h]$, and we again refer to these as components of generation means.

QUESTION Write down the components of the generation means in the P_1, P_2, F_1, F_2, B_1 and B_2 generations.

ANSWER
$$\bar{P}_1 = m + [d]$$
$$\bar{P}_2 = m - [d]$$
$$\bar{F}_1 = m + [h]$$
$$\bar{F}_2 = m + \tfrac{1}{2}[h]$$
$$\bar{B}_1 = m + \tfrac{1}{2}[d] + \tfrac{1}{2}[h]$$
$$\bar{B}_2 = m - \tfrac{1}{2}[d] + \tfrac{1}{2}[h]$$

11.4.4 Gene dispersion

When all the alleles of like effect occur in the same parental line, we refer to this situation as *complete association* (Section 11.4.1). Let us consider an example.

$$AA\ BB\ CC\ DD \qquad \text{genotype 1}$$
$$aa\ bb\ cc\ dd \qquad \text{genotype 2}$$

In genotype 1 all the increasing alleles of the genes are associated, and in genotype 2, all the decreasing alleles of the genes are associated.

Consider a third genotype:

$$AA \quad BB \quad cc \quad dd$$
$$+d_a \quad +d_b \quad -d_c \quad -d_d$$

Here, there are two genes of increasing effect and two of decreasing effect, and there is clearly no longer a state of complete association. In fact, because equal numbers of increasing and decreasing alleles are present in the same genotype, we refer to this situation as the *maximum dispersion* of genes.

gene dispersion

QUESTION Is there complete association or maximum dispersion of genes in the following pairs of genotypes? For ease of presentation we are using plus ($+$) to indicate increasing alleles and minus ($-$) to indicate decreasing alleles.

$$\begin{array}{l} P_1 \quad +\ +\ +\ +\ +\ + \\ P_2 \quad -\ -\ -\ -\ -\ - \end{array} \quad \text{parental pair 1}$$

$$\begin{array}{l} P_1 \quad +\ -\ +\ +\ -\ - \\ P_2 \quad -\ +\ -\ -\ +\ + \end{array} \quad \text{parental pair 2}$$

$$\begin{array}{l} P_1 \quad +\ +\ +\ +\ -\ - \\ P_2 \quad -\ -\ -\ -\ +\ + \end{array} \quad \text{parental pair 3}$$

ANSWER *Parental pair 1* Complete association, because all the increasing alleles occur together in one strain, and the decreasing alleles in the other.

Parental pair 2 Maximum dispersion, because equal numbers of increasing and decreasing alleles occur together in both strains.

Parental pair 3 Neither complete association nor maximum dispersion. In this parental pair, there is partial association or partial dispersion. The amount, or the degree of association, however, is very important in interpreting the meaning of [d], particularly when we come to discuss heterosis in Section 11.7.

We shall now see how to calculate the degree of association. In general, we can say that $[d] = r_d \sum (d)$, where r_d is the degree of association of the genes, and $\sum (d)$ is the sum of all the individual additive effects.

> QUESTION In parental pair 1 of the previous QUESTION and ANSWER, all the genes of like effect were associated in one parental strain and, therefore, $r_d = 1$. What is the value of [d]? (You may assume that each additive effect is equal to 1.)
>
> ANSWER $$[d] = r_d \sum (d) = 1 \times 6 = 6$$

If the genes are completely associated, $[d] = \sum (d)$, that is, [d] is equal to the sum of all the individual additive effects; therefore, $r_d = 1$ for complete association.

> QUESTION In parental pair 2 of the same QUESTION and ANSWER, the genes of increasing and decreasing effect were equally distributed between the two parental strains. This situation of maximum dispersion results in r_d being equal to zero. Calculate [d]. (Again, assume that each additive effect is equal to 1.)
>
> ANSWER $[d] = r_d \sum (d) = 0 \times 6 = 0$ because $\sum (d)$ still equals 6 and $r_d = 0$ for maximum dispersion.

With maximum dispersion, [d] does not equal $\sum (d)$. Now this distinction between $[d] = \sum (d)$ for complete association, and $[d] = 0$ for maximum dispersion, is very important. In Section 11.4.5 we shall discuss its implications when we reconsider the degree of dominance.

Of course, it would be unrealistic to think only in terms of the complete association or the maximum dispersion of genes, and we must now consider how r_d may be calculated in other situations.

For simplicity, we shall assume that two lines, P_1 and P_2, differ at K loci, and that the additive effects of each locus are equal, that is, $d_a = d_b = d_c = \cdots d_k$ (where d_k is the additive effect of the Kth locus).

Let us say that the P_2 parent has k' genes of increasing effect and, therefore, $K - k'$ genes of decreasing effect. As the two lines, P_1 and P_2, are homozygous for different alleles of the same genes, P_1 will have $K - k'$ genes of increasing effect and k' genes of decreasing effect.

Table 3

Parent	Increasing genes	Decreasing genes	Total number of genes
P_1	$K - k'$	k'	K
P_2	k'	$K - k'$	K

Now let us consider the P_1 parent. We know that

$$[d] = \sum (d_+) - \sum (d_-)$$

As

$$\sum (d_+) = (K - k')d$$

and

$$\sum (d_-) = k'd$$

so,

$$[d] = (K - k')d - (k')d$$
$$[d] = (K - 2k')d$$

Now, $K - 2k'$ is the balance effect between the increasing and decreasing alleles of the genes. If we express this relationship as a proportion of the total number of genes, that is, K, then $\dfrac{K - 2k'}{K}$ measures the degree of association of genes, which we have denoted by r_d.

We can now say that $[d] = r_d Kd$, where r_d is the degree of association of the genes and Kd is the additive effect of all K genes. You will recall that we assumed that all additive effects of the K genes were equal; therefore, the total additive effects are equal to $K \times d$, that is, Kd.

If we now relax the special condition that $d_a = d_b = d_c = \cdots d_k$, then $[d] = r_d \sum (d)$, where $\sum (d)$ is the sum of all the independent additive effects.

Let us examine the relationship between $[d]$ and $r_d \sum (d)$, by considering the parental lines $AA\ BB\ CC\ DD\ ee$ and $aa\ bb\ cc\ dd\ EE$. Again, for the sake of simplicity, assume that $d_a = d_b = d_c = d_d = d_e = 1$.

Consider the P_1 parent, $AA\ BB\ CC\ DD\ ee$.

The number of increasing alleles $= K - k' = 4$

The number of decreasing alleles $= \qquad k' = 1$

Total number of genes $\qquad = K \qquad = 5$

$$r_d = \frac{K - 2k'}{K} = \frac{5 - 2}{5} = \frac{3}{5}$$

$[d] = r_d \sum (d)$. $\qquad \sum (d) = 5$ and $r_d = \frac{3}{5}$, therefore, $[d] = \frac{3}{5} \times 5 = 3$.

We can also calculate $[d]$ from the relationship $[d] = \sum (d_+) - \sum (d_-)$.

QUESTION Calculate $[d]$ using the P_1 parent, from the relationship $[d] = \sum (d_+) - \sum (d_-)$.

ANSWER

	(d_+)	(d_-)	$[d]$
$AA\ BB\ CC\ DD\ ee$	4	1	3

Thus, $[d]$ is again equal to 3.

The significance of r_d may not be immediately obvious to you, but you should see at this stage that the additive genetic effect is made up of two independent parts, one of which reflects the distribution of alleles and the other their total effect. In Section 11.7.3, you will appreciate how an understanding of gene association and gene dispersion can be the key to grasping important genetic phenomena such as heterosis.

QUESTION Calculate r_d in the following pairs of genotypes. (You may assume equal additive effects, so that $d_k = 1$.)

$P_1 \quad AA\ BB\ CC\ DD\ ee\ \ FF$
$P_2 \quad aa\ \ bb\ \ cc\ \ dd\ \ EE\ ff$ $\qquad$ parental pair 1

$P_1 \quad AA\ bb\ \ CC\ dd\ \ EE\ ff$
$P_2 \quad aa\ \ BB\ cc\ \ DD\ ee\ \ FF$ $\qquad$ parental pair 2

ANSWER

parental pair 1

$$K - k' = 5$$
$$k' = 1$$
$$K \qquad = 6$$
$$r_d = \frac{K - 2k'}{K} = \frac{6 - 2}{6} = \frac{2}{3}$$

parental pair 2

$$K - k' = 3$$
$$k' = 3$$
$$K \qquad = 6$$
$$r_d = \frac{K - 2k'}{K} = \frac{6 - 6}{6} = 0$$

and, as we saw earlier, with maximum dispersion, r_d, and therefore $[d]$, are equal to 0.

In the same way that we defined $[d]$ as being equal to $r_d \sum (d)$, we now need to consider the effects of gene association on the heterozygous F_1.

The F_1 cross between two parental lines that differ in being homozygous for different alleles at the K loci must be heterozygous at all K loci irrespective of the distribution of the alleles between the two parents. Consequently, $[h]$ is independent of the degree of association of the genes and, therefore $[h] = \sum (h)$.

We can see, therefore, that $[h]$ and $[d]$ and $\sum (h)$ and $\sum (d)$ can be very different: $\sum (d)$ is the *total* additive effect of the genes and is equal to $[d]$ only when the degree of association r_d is equal to 1. On the other hand, $\sum (h)$ and $[h]$ are exactly the same: both are the balance between the dominance effects of increasing and decreasing alleles. *Only* when all the h effects have the same sign, that is, when there is uni-directional dominance, do $[h]$ and $\sum (h)$ represent the *total* dominance effect of the genes.

11.4.5 The degree of dominance and the potence ratio

You will recall from Section 11.3.4, that for a difference in a single gene, the ratio $\frac{h}{d}$ was used as a measure of the degree of dominance of the gene.

When this expression is used as a measure of the degree of dominance for more than one gene, we run into the problems of the direction of dominance and the degree of association of genes in the parental lines.

Let us spell out these problems. For a single gene-difference, the degree of dominance is given by $\frac{h}{d}$. The equivalent expression for a many-gene difference is given by:

$$\frac{\text{total dominance effects}}{\text{total additive effects}}$$

But, all we can measure are $[d]$ and $[h]$, which are not always equivalent to the total additive and dominance effects of the genes.

$[d]$ is equal to the total additive effect only when the degree of association is equal to 1, that is, when $r_d = 1$.

Similarly, $[h]$ is equal to the total dominance effect only when all the independent h effects have the same sign, that is, when there is unidirectional dominance.

As neither of these assumptions is likely to be true in general, the ratio $\frac{[h]}{[d]}$ cannot be regarded as a measure of dominance in practice, and it is therefore referred to as the potence ratio or the relative potence of the parents. If, for example, we have dominance in different directions, it is possible to obtain a relative potence of zero, even though the individual genes are showing dominance. Similarly, a relative potence of ∞ will result from maximum dispersion irrespective of dominance.

potence ratio

ITQ 6 *Assertion* The potence ratio is a measure of the degree of dominance.

Reason The use of $\frac{[h]}{[d]}$ as a measure of the degree of dominance is independent of degree of association of the genes, and the direction of dominance of the individual genes.

Pick the correct statement from (i)–(iv).

(i) Both the reason and the assertion are true.

(ii) The reason is true but the assertion is false.

(iii) The reason is false but the assertion is true.

(iv) Both the assertion and the reason are false.

ITQ 7 The mean yield of seed from two true-breeding lines of barley and their F_1 heterozygote are given below. (a) Calculate m, $[d]$, $[h]$ and the potence ratio. (b) Using your estimates of these parameters, predict the mean of the F_2 generation.

$$\overline{P}_1 = 355 \text{ g}$$
$$\overline{P}_2 = 135 \text{ g}$$
$$\overline{F}_1 = 300 \text{ g}$$

You should now attempt SAQs 2 and 3 (p. 549).

11.5 Scaling tests

An alternative heading for this Section would be 'The proof of the pudding', because we now need to test the adequacy of the models that we have been building.

We have explained the metrical relationship between pairs of genotypes in terms of the additive and dominance properties of the genes. But is this realistic? Is the relationship really meaningful?

In Sections 11.3.5 and 11.4.3, the components of generation means were derived for the six most common generations. Using these expectations we can actually test whether or not the metrical relationship between genotypes is adequately described in terms of d or $[d]$, and h or $[h]$ effects of the genes.

For a single-gene difference, for example the A–a gene, the following relationship between the generation means will be true within the limits set by sampling error:

$$2\overline{B}_1 - \overline{F}_1 - \overline{P}_1 = 0$$

where $\overline{B}_1$ = the mean of the B_1 generation

$\overline{F}_1$ = the mean of the F_1 generation

$\overline{P}_1$ = the mean of the P_1 generation

Such a relationship has been called a '*scaling test*', and there are in fact several different scaling tests available. In this Unit, however, we shall restrict ourselves to the use of the relationship shown above, which is identified as an A-scaling test.

scaling tests

$A = 2\overline{B}_1 - \overline{F}_1 - \overline{P}_1 = 0$. In other words, the value of A will be zero within the limits set by sampling error.

The parameters we are using are valid only if a randomized plot design has been used for the different generations.

QUESTION Using the components of generation means, that is, m, d and h, can you show why A should be equal to zero?

ANSWER
$$2\overline{B}_1 = 2(m + \tfrac{1}{2}d + \tfrac{1}{2}h)$$
$$\overline{F}_1 = m + h$$
$$\overline{P}_1 = m + d$$

Thus,

$$2(m + \tfrac{1}{2}d + \tfrac{1}{2}h) - (m + h) - (m + d) = 0$$

In terms of many-gene differences, it can be seen that

$$A = 2(m + \tfrac{1}{2}[d] + \tfrac{1}{2}[h]) - (m + [h]) - (m + [d]) = 0$$

Consider an example to illustrate the use of an A-scaling test.

The ear length in a variety of barley is as follows in the various generations:

$$\overline{P}_1 = 19 \text{ cm}$$
$$\overline{P}_2 = 12 \text{ cm}$$
$$\overline{F}_1 = 18 \text{ cm}$$
$$\overline{B}_1 = 18.5 \text{ cm}$$

Now, from $\overline{P}_1$, $\overline{P}_2$ and $\overline{F}_1$, we can calculate the parameters m, $[d]$ and $[h]$.

QUESTION Calculate m, $[d]$ and $[h]$.

ANSWER
$$m = \frac{\overline{P}_1 + \overline{P}_2}{2} = 15.5$$
$$[d] = \overline{P}_1 - m = 3.5$$
$$[h] = \overline{F}_1 - m = 2.5$$

Now, let us see if an additive dominance model is adequate.
$$A = 2(18.5) - (18) - (19) = 0$$

In this case, A is equal to zero and, therefore, an additive dominance model is an adequate description of the relationship between the genotypes.

QUESTION Below we give data for the mean number of grains per ear in true-breeding lines of barley and successive generations derived from a cross between them.

$$\overline{P}_1 = 9.125$$
$$\overline{P}_2 = 5.475$$
$$\overline{F}_1 = 5.500$$
$$\overline{B}_1 = 7.500$$

Calculate m, $[d]$ and $[h]$. Is an additive dominance model an adequate description of the metrical relationship between genotypes?

ANSWER
$$m = 7.300$$
$$[d] = 1.825$$
$$[h] = -1.800$$
$$A = 2(7.500) - (5.500) - (9.125) = 0.375$$

An A value of 0.375 may indicate that an additive dominance model is inadequate, or it may not. Before we can be really sure, the statistical significance of A must be assessed. In other words, we must assess whether A is significantly different from zero within the limits imposed by sampling error.

11.6 The statistical validation of scaling tests

You are strongly advised to read Section ST.6 of STATS before continuing with Unit 11.

You should recall from *STATS*, Section ST.7.5 that whether or not the deviation of a mean value estimated from a sample is significantly different from zero, or for that matter any other expected value, can be assessed by use of the *t*-test. Thus

$$t = \frac{\overline{x} - \mu}{\sqrt{\left(\dfrac{s^2}{n}\right)}}$$

where $\overline{x} =$ the estimate of the mean

$\mu =$ the expected value (in the case of A, $\mu = 0$), and

$\sqrt{\left(\dfrac{s^2}{n}\right)} =$ the standard error of the mean, SE

Using a *t*-test, therefore, it is possible to assess the significance of A.

To explain the use of the *t*-test adequately, we need to recapitulate to some extent. The standard error of a mean is given by:

$$\text{SE} = \sqrt{\left(\frac{s^2}{n}\right)} = \frac{s}{\sqrt{n}}$$

where s^2 is the variance of the individuals

s is the standard deviation of the individuals and

n is the number of individuals

We can think of the standard error in terms of the square root of the variance of the sampling distribution of the mean ($STATS$, Section ST.6.3).

To illustrate these points, consider the following data. The height of a true-breeding line of barley has been measured, and the mean of 36 individuals is found to be 100 cm with a variance of 25 cm^2.

QUESTION Calculate the standard error of the mean.

ANSWER

$$SE = \sqrt{\left(\frac{s^2}{n}\right)} = \sqrt{\left(\frac{25}{36}\right)} = \frac{5}{6} = 0.83$$

or

$$SE = \frac{s}{\sqrt{n}} = \frac{5}{\sqrt{36}} = \frac{5}{6} = 0.83$$

Thus, as the standard error is the square root of the variance about the mean, we can find the variance of the mean from $\frac{s^2}{n}$, and the square root of this from $\sqrt{\left(\frac{s^2}{n}\right)}$. In this example, therefore, $\frac{s^2}{n} = \frac{25}{36} = 0.694$ and SE $= \sqrt{0.694} = 0.83$.

From now on, when the mean of any generation is given, the standard error will also be attached unless stated otherwise. Thus:

$$\bar{P}_1 = 100 \pm 0.83$$

How do we calculate the standard error of A?

As

$$A = 2\bar{B}_1 - \bar{F}_1 - \bar{P}_1$$

you should recall from $STATS$, Section ST.7.2, that

$$V(A) = 4V(\bar{B}_1) + V(\bar{F}_1) + V(\bar{P}_1)$$

where $V(A) =$ the variance of the mean of A

$V(\bar{B}_1) =$ the variance of the mean of the B_1 generation

$V(\bar{F}_1) =$ the variance of the mean of the F_1 generation

$V(\bar{P}_1) =$ the variance of the mean of the P_1 generation

The standard error of A is then given by $\sqrt{V(A)}$.

The variances of the means of the B_1, F_1 and P_1 generations are obtained by squaring their respective standard errors. This follows because SE $= \sqrt{\left(\frac{s^2}{n}\right)}$, and $\frac{s^2}{n}$, that is, the variance of the mean $=$ SE^2.

Let us reconsider the data to which we applied an A-scaling test and found the value of A to be 0.375 (see p. 531).

This time we have added the standard error to the estimates of each of the three generation means, together with their degrees of freedom (one less than the number of individuals making up the estimate) (see right).

	Degrees of freedom $(n-1)$
$\bar{P}_1 = 9.125 \pm 0.091$	9
$\bar{F}_1 = 5.500 \pm 0.086$	9
$\bar{B}_1 = 7.500 \pm 0.100$	9

QUESTION Calculate the standard error of A (Remember A is equal to 0.375.) How would you show that A is significant?

ANSWER The variance of the mean of the three generations is obtained by squaring the appropriate standard error. Thus,

$$V(A) = 4V(\bar{B}_1) + V(\bar{F}_1) + V(\bar{P}_1)$$
$$= 4(0.100)^2 + (0.086)^2 + (0.091)^2$$
$$= 4(0.010\ 0) + (0.007\ 4) + (0.008\ 3)$$
$$= 0.055\ 7$$
$$SE(A) = \sqrt{0.055\ 7} = 0.235$$
$$A = 0.375 \pm 0.235$$

To test the significance of A, we employ the usual t-test

$$t = \frac{\bar{x} - \mu}{SE} = \frac{A - 0}{SE(A)} = \frac{0.375}{0.235} = 1.59$$

QUESTION In a normal t-test the degrees of freedom are $n - 1$, where n is the number of observations making up the estimate. Can you suggest how to calculate the degrees of freedom in an estimate of A?

ANSWER For A, we must add the degrees of freedom of the variance of each generation mean making up the estimate of A. Thus,

the degrees of freedom of $A = $ d.f. of $(\bar{B}_1) + (\bar{F}_1) + (\bar{P}_1)$

$$= 9 + 9 + 9$$
$$= 27.$$

So we are looking at a t-test with 27 degrees of freedom.

Reference to *STATS*, Table 3 on p. 57, shows quite clearly that with a t-value of 1.59 A is not significantly different from 0 at a level of probability of more than 0.10 (d.f. = 27).

We can conclude, therefore, that an additive dominance model is an adequate description of the metrical relationship between the genotypes.

ITQ 8 The following are data for the mean yield of seed of two true-breeding lines of barley, and successive generations derived from an initial cross between them. (a) Calculate the parameters m, $[d]$, $[h]$ and the potence ratio. (b) Is an additive dominance model an adequate description of the metrical relationship between genotypes?

Generation	Estimate of mean	Variance s^2	Number in sample
P_1	9.028	0.071 0	20
F_1	6.329	0.121 0	20
B_1	7.034	0.262 0	20
P_2	6.318	0.065 0	20

From ITQ 8 you should have found that the value of A was significantly different from 0 on a t-test with 57 degrees of freedom. What does this mean? Such a result tells us that an additive dominance model is an inadequate description of the metrical relationship between genotypes—so where do we go from here? Before we can reasonably discuss this, two important questions must be answered: first, what genetic phenomena could be responsible for the inadequacy of an additive dominance model and, second, what can we do with data that do not fit an additive dominance model?

It is impossible to discuss all the factors that might contribute to the inadequacy of an additive dominance model. For illustrative purposes, however, three important phenomena will be briefly outlined.

1 Abnormal chromosomal behaviour If the genetic contribution to gametes is unequal, then the frequency of genotypes, and hence their contributions in terms of m, $[d]$ and $[h]$ to the generation means, would not be as predicted, by Mendelian hypothesis.

2 Cytoplasmic inheritance If a character is determined by, or affected by, non-nuclear factors, then the expression of the character cannot be explained solely in terms of m, $[d]$ and $[h]$.

3 Non-allelic interactions or the effects of epistasis When more than one gene determines the expression of a character, then genes at *different* loci may affect one another's expression. For example, consider the AA and BB genes. Independently, each gene contributes $+d_a$ and $+d_b$ to the expression of a character. Thus:

$$AA \qquad BB$$
$$+d_a \qquad +d_b$$

When both these genes occur together in the same genotype, we assume that their additive effect on the expression of the character is the sum of their independent additive effects, that is, $+d_a + d_b = \sum(d_+)$. But what if this is not correct? There may well be an interaction that means that the final expression of $AABB$ is *not* equal to $+d_a + d_b$. In these circumstances, therefore, the expectations of generation means cannot be explained solely in terms of m, $[d]$ and $[h]$.

These three phenomena have been selected because they illustrate an important biological concept. A biometrical model is not an abstract quantitative description; it is a description of quantitative variation in terms of likely genetic phenomena, the simplest of these being additivity and dominance. Where an additive dominance model is inadequate, then other parameters may be built into the model to explain other sources of variation. Before we do this, however, it is only common sense to look for other less complicated explanations.

One example of such an explanation is abnormal chromosomal behaviour. During this Course you have been introduced to the methods of chromosome analysis. Using such techniques, meiotic preparations of the experimental material in question could be analysed. If the behaviour of the chromosomes was found to be abnormal, this alone could lead to the failure of an additive dominance model.

As biometrical analysis depends on the occurrence of genotypes with predictable frequency, any material that does not comply with normal functional diploid inheritance cannot be analysed satisfactorily by biometrical methods.

Similarly, if the expression of a character is determined or influenced by cytoplasmic factors, for example, maternal effects, the simple biometrical analysis is not applicable. For example, if the expression of the offspring from a cross between two hermaphrodite plants depends on from which of the two parents the offspring developed, then the F_1 generation mean cannot be explained solely in terms of m, and $[h]$.

There are complete biometrical genetic analyses that take into account maternal/paternal effects, etc.; they present no difficulties. The main point is that a biometrical analysis is only one of several approaches in analysing biological systems. No one approach is ever independent and it is, therefore, essential to consider the whole biology of an organism in any analysis.

Let us assume that simple explanations, such as non-diploid inheritance, have been ruled out by appropriate observations. We are now forced to add to the complexity of the m, $[d]$ and $[h]$ model by the introduction of new parameters.

Consider the case in which the difference between AA and aa is dependent on whether BB or bb is present in the genotype. Let

$$AABB = 24$$
$$AAbb = 12$$
$$aaBB = 12$$
$$aabb = 8$$

The difference between AA and aa when BB is present is

$$24 - 12 = 12$$

and when bb is present, the difference is

$$12 - 8 = 4$$

The effect of substituting BB for bb increases the additive effect of the $A–a$ gene. The action of the alleles at one locus, therefore, is not independent of the action of the alleles at the other locus. Interaction between non-allelic genes (or *epistasis*) is, therefore, occurring.

epistasis

We can say that there is a homozygote × homozygote interaction between the additive effects at the A and B loci. This interaction is symbolized by the parameter i, and because it is an interaction between d_a and d_b, the i is specified by i_{ab}.

There is no need to discuss this in any further detail, but it is important to realize that interactions may occur between different genes, and that the contribution made by their respective parameters may be predicted in different generation means.

It is, therefore, possible to quantify the metrical relationship between genotypes, taking into account such interactions. To these ends, the majority of practical biometrical geneticists usually use a 'joint scaling test', sometimes called a *Cavalli analysis*, in which as many parameters as necessary may be quantified and their significance assessed in one operation. However, such a test depends on mathematical principles with which we cannot assume you will be familiar.

Cavalli analysis

To return to our original question of where to go, or what to do when an *A*-scaling test indicates the inadequacy of an additive dominance model—we can only say that the data must be ignored as we cannot consider such interactions in depth here. Therefore, for the remainder of Unit 11 and all of Unit 12, we shall use data in which an additive dominance model is an adequate description of the metrical relationship between genotypes.

Before we leave this Section, however, one more important concept requires detailed discussion—the concept of scale of measurement.

Let us examine a situation in which a character has been measured in a scale of centimetres and the effects of epistasis have been found.

	Genotypes		Difference between *AA* and *aa*
aabb 1	*Aabb* 2	*AAbb* 3	2
aaBb 2	*AaBb* 4	*AABb* 6	4
aaBB 3	*AaBB* 6	*AABB* 9	6

The expression of the *A–a* gene is dependent on which allele of the *B–b* gene is present within the genotype, and there is no dominance. Analysis of this data clearly shows that the effects of the genes, when measured in centimetres, are not linear or additive, but are multiplicative, where

$$aa = bb = 1$$
$$Aa = Bb = 2$$
$$AA = BB = 3$$

so *aabb* = 1 × 1 and *aaBB* = 1 × 3, etc.

If this data is transformed into $\log_{10}$ centimetres, the effects of epistasis can be removed, but dominance is now present.

	Genotypes		Difference between *AA* and *aa*
aabb 0	*Aabb* 0.301	*AAbb* 0.477	0.477
aaBb 0.301	*AaBb* 0.602	*AABb* 0.778	0.477
aaBB 0.477	*AaBB* 0.778	*AABB* 0.954	0.477

Thus, by transforming the scale of measurement into $\log_{10}$ cm, the effects of interaction can be converted into dominance effects, and this data can then be analysed satisfactorily on an additive dominance model.

To be quite truthful, these data are purely for demonstration purposes. Only occasionally will a change of scale remove epistatic effects and so allow the analysis to proceed on an additive dominance model. When no change of scale has the desired effect, we cannot proceed with the simple analysis.

The use of a different scale in no way invalidates the genetic analysis. What must be appreciated, however, is that the conclusions relate *only* to the character as measured on that particular scale. A change of scale will change the results and, therefore, the conclusions, and, of course, any prediction or extrapolation made from the conclusions will apply only to measurements made on the same scale.

Let us illustrate the consequences of a change of scale by referring to a simple example. Consider the *A–a* gene that determines seed length in a variety of bean, where *AA* is a tall plant and *aa* is a small plant. As our rulers are marked in centimetres, we measure the beans according to this scale and find that the mean of the *AA* genotypes is 9 cm, the mean of the *aa* genotypes is 1 cm, and the mean of the F_1 genotypes , *Aa*, is 4 cm.

> QUESTION Calculate m, d_a and h_a. Which allele of the *A–a* gene is showing incomplete dominance?

ANSWER

$$m = \frac{9 + 1}{2} = 5$$

$$d_a = 9 - 5 = \quad 4$$
$$h_a = 4 - 5 = -1$$

Thus, the a allele is showing incomplete dominance.

QUESTION This time we decide to express the length of the beans in $\log_{10}$ cm, and find that $AA = 0.93$, $Aa = 0.60$ and $aa = 0.00$. Calculate m, d_a and h_a. Which allele of the A–a gene is showing incomplete dominance?

ANSWER

$$m = \frac{0.93 + 0.0}{2} = \quad 0.465$$

$$d_a = 0.93 - 0.465 = \quad 0.465$$
$$h_a = 0.60 - 0.465 = +0.135$$

and the A allele is showing incomplete dominance. Thus, by changing the scale of measurement, the dominance relationship between the alleles of the A–a gene has been changed.

We have, therefore, demonstrated that the genetic analysis of data relates only to the scale on which the character was measured.

For the remainder of Unit 11 we shall leave the essential but, nevertheless, theoretical, consideration of biometrical genetics, and discuss its role as an analytical tool in the elucidation of an important plant and animal breeding phenomenon, which is termed hybrid vigour, or heterosis.

You should now attempt SAQ 4 (p. 549).

11.7 Hybrid vigour—heterosis

In this Section and in Section 11.8, we shall examine how a working knowledge of biometrical genetics can help to explain why hybrid strains tend to have more vigorous growth than the parental strains—that is, they display heterosis (hybrid vigour). We shall consider the more theoretical aspects of this subject in the text, and in the TV programme (Heterosis) the practical problems of breeding heterotic F_1 hybrid varieties will be illustrated with reference to maize and brussels sprouts.

There are many excellent textbooks on genetics but, unfortunately, each author appears to adopt his own definition of heterosis. In this Unit, therefore, we shall define heterosis in a way that is acceptable to most geneticists.

When naturally outbreeding plants and animal species are inbred over several generations, the phenomenon of *inbreeding depression* occurs. This can be recognized by a loss of overall fitness, a lowering of yield, a reduction in the numbers of seeds produced and a reduced ability to compete.

inbreeding depression

If two such inbred lines are crossed, however, the resulting F_1 hybrid may recover much of the lost vigour and, indeed, can be more vigorous than either of the two inbred parents. The extent to which such an F_1 generation is superior even to the better of its parents is termed heterosis. *Heterosis* can therefore be defined as the difference between the F_1 generation mean and the mean of the better of its parents. But what do we mean by the better parent? In the instances of the seed yield in cereals or the rate of food conversion in animals, for heterosis to occur the mean of the F_1 generation must be larger than the mean of the P_1 parent. That is: $\bar{F}_1 > \bar{P}_1$. (You should recall that the parent with the higher score is termed P_1 and the parent with the lower score is termed P_2.) If, on the other hand, we consider a character such as early ripening in cereals, then, for heterosis to occur, the mean of the F_1 generation must be less than the mean of the P_2 parent. That is: $\bar{F}_1 < \bar{P}_2$.

heterosis

To understand the importance of inbreeding and heterosis, we must appreciate both the genetic causes and the socio-economic reasons for its wide application in plant and animal breeding.

11.7.1   The application of inbreeding and heterosis to breeding plants and animals

Modern agricultural methods and, indeed, commercial requirements, consumer preferences and government legislation dictate the size, weight and quality of many of our food sources. For example, the deep-freeze industry requires brussels sprouts, chickens, ducks and many other foods to fall within very precise ranges of size and quality. Such uniformity can be obtained only by using lines that are as much as possible genetically identical.

Similarly, when agricultural labour was relatively inexpensive, the harvesting of many crop plants was done by hand. When, however, agricultural wages began to approach industrial wages, manual labour was no longer economic and mechanical harvesting was the only viable alternative.

Look at the following figures that show the yield of sprouts per hectare in an outbreeding variety of brussels sprout:

manual harvesting	$c.$ 3.2 tonnes per hectare
mechanical harvesting	$c.$ 1.2 tonnes per hectare

When sprouts were harvested by hand, only the ripe sprouts were taken. As any outbreeding variety consists of genetically diverse individuals, different plants ripened at different times, and this entailed several visits to each plant. This system was clearly the most efficient in achieving a maximum crop, but was uneconomical in terms of manpower. On the other hand, mechanical harvesting merely strips off every plant, regardless of whether or not the sprouts are ready, and consequently, the overall marketable yield is substantially reduced.

Now look at the yield of an F_1 hybrid variety of sprout:

mechanical harvesting	$c.$ 3.64 tonnes per hectare

The uniformity of the F_1 hybrid, together with the economies of mechanical harvesting, have resulted in optimum productivity. There is, therefore, a pressing demand for *uniform* varieties of crop plant species.

A further advantage that has been gained from brussels sprout breeding is that the latest varieties (1975) feature uniform ripening of sprouts on individual plants rather than the more usual sequential ripening from the bottom to the top of the plant; this increases the optimum yield even further.

> QUESTION Which of the six generations you are familiar with are genetically uniform?
>
> ANSWER Inbred lines of plants and animals will be genetically uniform. The F_1 hybrid from two different inbred lines will also be genetically uniform, but all other generations will undergo segregation and, therefore, be genetically heterogeneous.

If we are to produce genetically uniform, outbreeding species of crop plants and domestic animals, then F_1 hybrids and inbred lines appear to be the alternative methods. But which of the two is the more practicable? (Of course, many crop plants are naturally inbreeding anyway, and in these species the problem does not arise.)

11.7.2   Inbred lines or F_1 hybrids?

When normally outbreeding species of plants and animals are inbred over several generations, inbreeding depression occurs at a rate that is dependent on the level of inbreeding. This inbreeding depression manifests itself in a deterioration in vigour and health and can be related to the rapid loss of heterozygosity and the fixation in the homozygous state of disadvantageous alleles, which in the normal heterozygous state are not expressed.

You should recall from Units 9 and 10, Section 9.1.1, that the degree of heterozygosity decreases by 50 per cent per generation of selfing, selfing being the ultimate in inbreeding.

QUESTION Starting with the heterozygote Aa, show that its frequency decreases by 50 per cent per generation of selfing. Calculate the generation mean for two generations of selfing.

ANSWER

	Genotypes					Generation mean
F$_1$ generation			Aa			
frequency, f			1			
mean			$m + h$			$m + h$
first generation of selfing	AA		Aa		aa	
frequency	$\frac{1}{4}$		$\frac{1}{2}$		$\frac{1}{4}$	
mean	$\frac{1}{4}(m+d)$		$\frac{1}{2}(m+h)$		$\frac{1}{4}(m-d)$	$m + \frac{1}{2}h$
second generation of selfing	AA	AA	Aa	aa	aa	
frequency	$\frac{1}{4}$	$\frac{1}{8}$	$\frac{1}{4}$	$\frac{1}{8}$	$\frac{1}{4}$	
mean	$\frac{1}{4}(m+d)$	$\frac{1}{8}(m+d)$	$\frac{1}{4}(m+h)$	$\frac{1}{8}(m-d)$	$\frac{1}{4}(m-d)$	$m + \frac{1}{4}h$

Thus the frequency of the heterozygotes decreases by 50 per cent per generation of selfing, and the generation mean decreases towards the value of m.

We can see that the mean of any generation, which we term $\overline{G}$, in this series of selfing is equal to

$$\overline{G} = m + fh,$$

and for many genes, $\quad\overline{G} = m + f[h]$

where f is the frequency of heterozygotes.

As the frequency of heterozygotes decreases on inbreeding, the cause of inbreeding depression must be directly related to the loss of heterozygosity and with it the dominance effects of the genes.

It would appear, therefore, that the potential use of inbred lines as a source of new and uniform varieties of crop plants and domestic animals has very serious limitations for outbreeding species or for characters showing directional dominance in such species. Any advantages conferred by genetic uniformity may be overshadowed by the detrimental effects of inbreeding depression.

Well, the picture is not quite as hopeless as it appears; but before we discuss inbred lines any further, we should consider the practical problems in breeding F$_1$ hybrids.

Let us consider the case in which $\overline{F}_1 > \overline{P}_1$. On crossing two inbred lines, the F$_1$ will be heterozygous at all loci at which the two homozygous lines were different. As the frequency of heterozygotes in the F$_1$ generation relative to their frequency in other generations is equal to 1, then from the relationship, $\overline{G} = m + f[h]$, it follows that the $\overline{F}_1$ will be greater than the mean of the inbred parental generation as a whole, provided that $[h] > 0$, that is, the h effects are predominantly unidirectional.

In other words, heterosis is caused by the restoration of the dominance effects of unidirectionally dominant genes.

But, of course, a plant or animal breeder does not compare the F$_1$ generation mean with the mean of the inbred generation as a whole, because this would serve no particular purpose. The practical aspects of breeding for heterosis will be discussed in Section 11.8. At this stage, however, it is essential to realize that breeders produce inbred lines and select among these for potential parents of their F$_1$ hybrids, thus reducing the frequency of disadvantageous alleles in the breeding material. Crosses are then made between pairs of selected inbred lines and the relative performances of the F$_1$ hybrids are assessed.

You will recall from the beginning of this Section, that heterosis was defined as the difference between the mean of an F$_1$ cross and the mean of the better of its parents. It can be defined, therefore, in terms of the mean of the parental generation, that is, $m + d$ or $m + [d]$, and the mean of the F$_1$ generation, that is, $m + h$ or $m + [h]$.

For the mean of the F_1 generation to be larger than the mean of the P_1 generation, then, for a single-gene difference $m + h$ must be greater than $m + d$, and for many gene differences $m + [h]$ must be greater than $m + [d]$. In other words, h must be positive and greater than d.

For the mean of the F_1 generation to be smaller than the P_2 generation, for a single gene difference $m + h$ must be greater than $m - d$, and for many gene differences $m + [h]$ must be greater than $m - [d]$. In other words, h must be negative and greater than d.

In order to understand heterosis we now need to consider the conditions under which h is greater than d.

11.7.3 The specification of heterosis

On a one-gene model, then, for $\overline{F}_1$ to be greater than $\overline{P}_1$, h_a must be positive and greater than d_a.

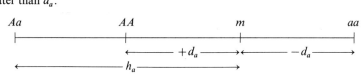

It follows that the degree of dominance, $\dfrac{h_a}{d_a} > +1$, that is, there is 'overdominance'.

Similarly, for $\overline{F}_1$ to be smaller than $\overline{P}_2$, h_a must be negative and greater than d_a. Once again, there is overdominance because the degree of dominance, $\dfrac{h_a}{d_a} > -1$.

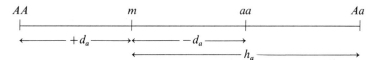

If we now consider the case in which differences in more than one gene determine a character, then for $\overline{F}_1$ to be greater than $\overline{P}_1$, $[h]$ must be positive and greater than $[d]$, and for $\overline{F}_1$ to be smaller than $\overline{P}_2$, $[h]$ must be negative and greater than $[d]$.

This could arise in one of two ways (we should bear in mind that $[h] = \sum (h)$ and $[d] = r_d \sum (d)$ and that h can take a positive or negative sign). First, $\dfrac{h}{d} > 1$ at one or more loci, that is, there is overdominance; second, $r_d < 1$, that is, there is gene dispersion.

Neither overdominance nor gene dispersion alone would give heterosis; they would need to be accompanied by unidirectional dominance, that is, by h effects of predominantly the same sign. If overdominance is the cause of heterosis, then the total dominance effect must be greater than the total additive effect. However, if gene dispersion is the cause, this is not a necessary condition. If $r_d < 1$, then the apparent or the observed additive effect $[d]$ will be reduced, and $\sum h$ will appear to be larger.

To measure the total values, we need two further parameters that are components of variance. At this stage, however, it is sufficient to know that where actual estimates of total dominance effects and total additive effects have been made, the degree of dominance has not been found to be greater than one in most instances. In other words, dispersed dominant genes are, in general, the cause of heterosis.

Let us now look at the way in which heterosis arises from the dispersal of dominant genes by considering in more detail two lines in which the genes are completely associated.

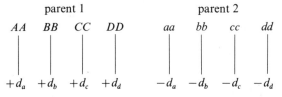

and the value of $[d] = r_d \sum (d)$. If we care to assume that the additive effect of each

gene is equal to 1 unit—that is, $+d_a = +d_b = +d_c = +d_d = 1$—then $[d] = 4$.
(This follows because $r_d = 1$ and $\sum(d) = 4$.)

Now consider the F_1 heterozygote from two such strains:

F$_1$ generation
Aa   Bb   Cc   Dd

h_a h_b h_c h_d

If we again assume that each h effect is equal to 1 unit, and that such effects are unidirectional, then

$$[h] = \sum(h) = +4$$

Therefore, $\overline{F}_1 = \overline{P}_1 = 4$, and heterosis is not occurring. To obtain heterosis in this example, we should require overdominance at one or more loci.

Now, consider two homozygous lines that share their genes equally and, therefore, show maximum dispersion.

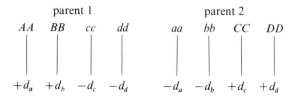

parent 1
AA   BB   cc   dd

$+d_a$ $+d_b$ $-d_c$ $-d_d$

parent 2
aa   bb   CC   DD

$-d_a$ $-d_b$ $+d_c$ $+d_d$

QUESTION   Calculate $[d]$, assuming that $+d_a = +d_b = +d_c = +d_d = 1$.

ANSWER $[d] = \sum(d_+) - \sum(d_-) = 2 - 2 = 0$

On crossing two such lines, the F_1 is heterozygous at all loci:

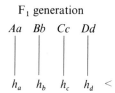

F$_1$ generation
Aa   Bb   Cc   Dd

h_a h_b h_c h_d $<$

Again, if we assume that $h_a = h_b = h_c = h_d = 1$, and all the h effects are unidirectional, then

$$[h] = \sum(h) = +4$$

Therefore, $[h] = 4$, and $[d] = 0$, so the F_1 hybrid is showing heterosis.

Let us examine the assumptions we have just made: first, that the genes were showing maximum dispersion and, second, that the direction of dominance was uniform at all loci.

Recalling that $[d] = r_d \sum(d)$, we can say that heterosis will occur where r_d is less than 1 and tends to maximum dispersion, that is $r_d = 0$, and, therefore, $[d] = 0$.

Similarly, $[h] = \sum(h)$. Providing that the genes are showing complete or incomplete dominance preponderantly in one direction, then where $[d]$ is less than $\sum[d]$, that is, $r_d < 1$, then $[h]$ can be greater than $[d]$ and heterosis can occur.

QUESTION   Consider the following pairs of inbred lines and determine whether or not the F_1 hybrids would be heterotic. You may assume that $+d_a = +d_b = +d_c = +d_d = 1$, and that $h_a = h_b = h_c = h_d = +1$

	parent 1	parent 2
parental pair 1	AA BB cc	aa bb CC
parental pair 2	aa bb CC DD	AA BB cc dd
parental pair 3	AA BB CC DD	aa bb cc dd

ANSWER

parental pair 1

$$AA \quad BB \quad cc \qquad aa \quad bb \quad CC$$
$$+d_a \quad +d_b \quad -d_c \qquad -d_a \quad -d_c \quad +d_c$$
$$[d] = \sum (d_+) - \sum (d_-) = 1$$

F_1 generation

$$Aa \quad Bb \quad Cc$$
$$h_a \quad h_b \quad h_c$$
$$[h] \sum (h) = +3$$

As $[h] = 3$, and $[d] = 1$, heterosis is occurring.

parental pair 2

$$aa \quad bb \quad CC \quad DD \qquad AA \quad BB \quad cc \quad dd$$
$$-d_a \quad -d_b \quad +d_c \quad +d_d \qquad +d_a \quad +d_b \quad -d_c \quad -d_d$$
$$[d] = \sum (d_+) - \sum (d_-) = 0$$

F_1 generation

$$Aa \quad Bb \quad Cc \quad Dd$$
$$h_a \quad h_b \quad h_c \quad h_d$$
$$[h] = \sum (h) = +4$$

As $[h] = 4$, and $[d] = 0$, heterosis is occurring.

parental pair 3

$$AA \quad BB \quad CC \quad DD \qquad aa \quad bb \quad cc \quad dd$$
$$+d_a \quad +d_b \quad +d_c \quad +d_d \qquad -d_a \quad -d_b \quad -d_c \quad -d_d$$
$$[d] = \sum (d_+) - \sum (d_-) = 4$$

F_1 generation

$$Aa \quad Bb \quad Cc \quad Dd$$
$$h_a \quad h_b \quad h_c \quad h_d$$
$$[h] = \sum (h) = +4$$

In parental pair 3 both $[d]$ and $[h]$ have the same value, so heterosis would not be possible unless there was overdominance at one or more loci.

To summarize: heterosis may be explained by overdominance at one or more loci, or by the dispersion of genes whose dominance effects are preponderantly unidirectional.

But does knowing the cause of heterosis serve any practical function? Will it help the plant and animal breeders to produce more food, cheaper food and better quality food? The answer to these questions is an overwhelming yes. However, from your knowledge of genetics, it should be obvious that one of the major problems is that the hybrid seed must be re-created from the two inbred parents every generation. Heterosis is a property of heterozygotes only. Once an F_1 generation is allowed to self, segregation takes place, resulting in progeny only half of which will be heterozygous at any one locus and, therefore, heterotic like the F_1; the other half will be the two types of homozygote that were used as the original parent strains.

However, if this F_1 heterosis were caused by dispersed dominant genes rather than by overdominance, then it should be possible to produce a pure-breeding strain in which all the increasing dominant genes are completely associated in the homozygous state. In this way a pure-breeding strain would be obtained having the same or a similar phenotypic expression to the original heterotic F_1, and we would have achieved the genetic uniformity of non-segregating generations without the deleterious effects of inbreeding depression. Likewise, the 'heterotic' effects would be fixed in a pure-breeding line, thus eliminating the necessity to re-create a hybrid seed stock each generation.

Before considering this most interesting concept, let us turn our attention to the practical problems and limitations of breeding F_1 hybrids.

You should now attempt SAQ 5 (p. 549).

11.8 Breeding F_1 hybrids

The production of heterotic F_1 hybrids involves three basic phases:

F_1 hybrids

1 Inbreeding a variety of parental lines and, from among the inbred parents, identifying potential parents. The actual degree of inbreeding will be dependent on the breeding system of the organism.

2 Producing numerous crosses between pairs of potential parents.

3 Testing each cross and selecting those from which the best F_1 hybrid is produced. Maximum heterosis requires combining dominant alleles dispersed among different individuals, thus leading to genotypes containing a high frequency of dominant increasing alleles, or dominant decreasing alleles if we wish $\overline{F}_1$ to be smaller than $\overline{P}_2$ (e.g. ripening time).

It is not possible to consider all the practical problems that are incurred when attempting to breed for heterotic F_1 hybrids. For the sake of simplicity, however, we have outlined two basic areas in which problems are known to arise: first, the production of inbred parental lines and, second, the production of the F_1 hybrid.

11.8.1 Practical limitations of breeding heterotic F_1 hybrids in plants

Consider a hermaphrodite plant that is capable of self-fertilization; the production of inbred lines in this case only entails ensuring that stray pollen does not fertilize the plants. This can be achieved relatively simply by covering the inflorescence with a suitable bag or by growing the plant in isolation. When two such inbred lines are hybridized, how can we be sure that the resulting seeds are F_1 and not parental types?

To ensure the production of F_1 seeds, each flower would have to be emasculated, that is, the anthers would have to be removed, and the pollen from the male parent would have to be transferred by hand to the stigma of the female parent. The time involved in this procedure would make such an operation very expensive.

On the other hand, some plants are self-incompatible (Units 9 and 10, Section 10.1.6) and in order to inbreed them, individual flowers must be tricked into accepting their own pollen and this has to be done by hand. The technique, called *bud pollination*, is demonstrated in the TV programme (Heterosis). When two such self-incompatible, inbred, lines are allowed to pollinate naturally, however, the likelihood of obtaining selfed progeny in the F_1 seed is very low.

> QUESTION   Can you see why the plant breeder is concerned about the proportion of parental types in his F_1 hybrid seed?
>
> ANSWER   As the mean of the parental generation is m and that of the F_1 generation is $m + [h]$, then, if the F_1 shows heterosis any number of parental types will reduce the mean yield of the generation overall, and the presence of parental types will also lead to genetic heterogeneity in the crop. Consequently, any F_1 hybrid variety that has a high proportion of parental types, is likely to be unpopular with the commercial growers and, therefore, a financial loss to the plant breeder.

As at least part of the hybridization procedure involves tedious 'man hours', the cost of the hybrid seed tends to be high. Also, the F_1 seed must be re-created every generation from parental lines that themselves are likely to be suffering from inbreeding depression and low fertility. The use of hybrid seed is, therefore, restricted to situations in which the advantages of F_1 hybrids in yield, etc., outweight the greater cost of seed production compared with other systems of plant breeding. However, many breeders are now exploiting genetic systems that facilitate hybrid formation, for example, male–sterility and self-incompatibility.

We shall now turn our attention to the possibility of fixing heterosis in pure-breeding lines. To understand this concept we need to compare the two major types of breeding systems in plants—inbreeding and outbreeding.

ITQ 9 Suppose you are a plant breeder who has been fortunate enough to produce a good heterotic F_1 hybrid of onion. The genotypes of your two inbred lines are

P_1 generation	P_2 generation
AA BB CC DD ee ff	*aa bb cc dd EE FF*

(a) What will be the genotype of the F_1 hybrid?

(b) Assuming that $+d_a = +d_b = \cdots +d_f = 1$, and that $h_a = h_b = \cdots h_f = +0.5$, calculate the values of $[d]$ and $[h]$.

So you now have a good heterotic variety, which you sell to the farmers and commercial growers. Because of your high production costs and capital outlay you are obliged to sell your seed at a high price. One farmer, who incidentally knew little of genetics, decided to allow his F_1 hybrid crop to set seed and then to use his own 'hybrid' seed the following year. In this way he expected to bulk up his own stock at no extra cost.

(c) What would the crop derived from the farmer's own 'hybrid' seed be like?

We now arrive at the economic–genetic subtlety of F_1 hybrids. The breeders have the inbred parental lines, and are very keen to ensure that both parental types never occur together in the same seed stock so that only they can re-create the F_1 hybrid for the open market. Irrespective of any legal protection offered by the Plant Breeders Rights Bill, they have a 'genetic copyright' to their own varieties. So, when a plant breeder collects the F_1 hybrid seed from the two parental lines, he collects the F_1 seed from the two parents separately. He then sells the two sources of F_1 seed to different parts of the country or different regions of the world.

(d) Can you see why he needs to do this?

11.8.2 Inbreeding and outbreeding species

When outbreeding species of plants and animals are inbred over several generations, we know that homozygosity and inbreeding depression occur at a rate that is dependent on the level of inbreeding. Inbreeding will result in the expression of all the recessive alleles present; in the more common heterozygous state these would not have been expressed because of the presence of the dominant alleles. This is referred to as the 'fixation' of the recessive alleles in the population; they will continue to be expressed unless outbreeding occurs and dominant alleles are introduced. However, there are naturally inbreeding species, and these are in stark contrast to the outbreeding species that have been inbred, in that they are fertile, vigorous and show no signs of inbreeding depression even though they are effectively homozygous at all loci. How can such a contrast be explained?

Outbreeding and naturally inbreeding species differ in that in the former the population consists of individuals that are heterozygous at most loci, whereas in the latter the population consists of effectively homozygous individuals. In both types, however, the individuals must be genetically well-balanced in the sense that they are vigorous, healthy and fit, having passed the tests imposed by natural selection pressures.

In a normally outbreeding species, the disadvantageous alleles occur in the heterozygous state, and, therefore, only when they become homozygous will selection have an opportunity to remove them from the gene pool.

In an inbreeding species, however, the disadvantageous alleles will very quickly become homozygous and be expressed because of the nature of the breeding system. Under this regime, any combination of genes that led to loss of fitness would be very soon removed from the gene pool by natural selection.

How does this information help us in our understanding of the fixing of heterosis?

Inbreeding species are both fit and homozygous. Homozygosity, therefore, does not always imply loss of fitness and, with this in mind, it should be possible to develop a pure-breeding line in which all the increasing or decreasing dominant alleles are fixed, that is, a line in which all the genes of like effect are completely associated and, therefore, $r_d = 1$. Providing that $\sum(d)$ is equal to or greater than $\sum(h)$, the pure-breeding homozygote will be equal to or greater than the F_1 hybrid.

QUESTION Under what condition would $\sum (d)$ not be equal to or greater than $\sum (h)$?

ANSWER If there were overdominance at one or more loci, $\sum (d)$ would not be equal to or greater than $\sum (h)$.

You might, of course, ask why a completely associated strain does not already exist in nature. Such a strain may well exist if it is as fit or fitter than other strains that are not completely associated. But there is a paradox about the difference between natural fitness and what a plant or animal breeder requires when he assesses strains for heterosis. Heterosis does not mean fitness in the same way in which it was defined in Units 9 and 10. It can mean fewer seeds, more leaves and even strains of plants and animals that could not survive in nature because they require intense husbandry and special nutrition. We give as an example the possible genotypes and their respective weights in a grass species in which seed weight is determined by two genes, A–a and B–b.

Gametes	AB	Ab	aB	ab
AB	$AABB$ 22 mg	$AABb$ 19 mg	$AaBB$ 19 mg	$AaBb$ 16 mg
Ab	$AABb$ 19 mg	$AAbb$ 12 mg	$AaBb$ 16 mg	$Aabb$ 9 mg
aB	$AaBB$ 19 mg	$AaBb$ 16 mg	$aaBB$ 12 mg	$aaBb$ 9 mg
ab	$AaBb$ 16 mg	$Aabb$ 9 mg	$aaBb$ 9 mg	$aabb$ 2 mg

The weights have been calculated from:

$$+d_a = +d_b = \quad 5 \text{ mg}$$
$$h_a = \quad h_b = +2 \text{ mg}$$
$$m = \quad\quad 12 \text{ mg}$$

Let us assume that this particular species is living in open pasture. The genotype $AABB$ is producing the most seed and would, therefore, appear to be at a selective advantage. But this weight of seed may be too heavy for the flower stalk to support, resulting in no seeds being set, and a fitness of zero.

The other homozygous genotype $aabb$ is producing only a small amount of seed and the likelihood that this seed will find suitable living space for its establishment and survival is small, and therefore this strain too is unfit. Ideally, there should be a balance between the total weight of seed and the carrying capacity of the inflorescence, and the chance of establishment and survival of individual seed.

In most habitats, selection is for *optimum* expression of character. Optimum occasionally is the same as maximum, depending on the selection pressures operating.

A plant breeder interested in producing a new variety that sets many seeds would probably fail to find the genotype $AABB$ growing wild because of its lack of fitness under natural conditions. This does not mean that given time such a genotype could not be produced in a breeding programme. Once in cultivation, the strength of the flower stalk could be increased by artificial selection or by an improvement in husbandry or even by the provision of shelter from the wind.

The optimum expression of character is not in general the same as the maximum expression of character in natural environments. The plant breeder is often interested in extremes of character expression. Modern techniques of crop-plant husbandry enable him to supply a suitable environment for genotypes that do show extremes of character expression.

QUESTION List the evidence that suggests that a pure-breeding line could be produced that has the same phenotypic value as a heterotic F_1 hybrid.

ANSWER First, in most cases the overdominance theory does not explain observed heterosis. This suggests that the superiority of the F_1 hybrid is due to the dispersion of dominant alleles in the parental lines. Second, as a normally inbreeding species does not suffer from inbreeding depression, it is a fact that a homozygous line of a normally outbreeding species can be produced that has the same level of expression as a heterotic F_1.

What conditions are necessary for a pure-breeding line to have the same level of character expression as a heterotic F_1? For heterosis to occur we know that $\overline{F}_1 > \overline{P}_1$. Therefore, for the parental generation mean to be equal to or greater than the F_1 generation mean, $[d]$ must be equal to or greater than $[h]$, that is, $r_d \sum (d)$ must be equal to or greater than $\sum (h)$.

Provided that the degree of association, r_d, is equal to or greater than $\dfrac{\sum (h)}{\sum (d)}$, the parental generation mean will be greater than the F_1 generation mean, and the 'heterosis' will be fixed in the homozygous state.

QUESTION Consider the heterozygote *Aa Bb Cc Dd*, where $h_a = h_b = h_c = h_d = +1$. Assuming that $+d_a = +d_b = +d_c = +d_d = 1$, what degree of association would be necessary in the parental generation to ensure that $\overline{P}_1$ was equal to $\overline{F}_1$?

ANSWER For $\overline{P}_1$ to be equal to $\overline{F}_1$, $[d]$ must be equal to $[h]$.

We know that $[h] = \sum (h) = 4$, so $[d]$ must be equal to 4.

The *total* additive effects of the four genes, *AA*, *BB*, *CC* and *DD* are equal to 4, that is, $\sum (d)$ is equal to 4.

As $[d] = r_d \sum (d)$, and $\sum (d)$ must equal 4, r_d must be equal to 1. In other words, there must be complete association of genes in the parental lines.

In this particular example $\sum (h) = r_d \sum (d)$, and the degree of association necessary for $[d]$ to be equal to $[h]$ is also shown to be 1.

Thus $r_d = \dfrac{4}{4} = 1$, and the P_1 would have the genotype: *AA BB CC DD*.

QUESTION Consider the heterozygote *Aa Bb Cc Dd Ee*, where $h_a = h_b = h_c = h_d = +1$, and $h_e = -1$. Assuming that $+d_a = +d_b = +d_c = +d_d = +d_e = 1$, what degree of association in the parental lines would be necessary for $\overline{P}_1$ to be equal to or greater than $\overline{F}_1$?

ANSWER For $\overline{P}_1$ to be equal to $\overline{F}_1$, $[d]$ must be equal to $[h]$. We know that

$$[h] = \sum (h) = [+1 +1 +1 +1 +(-1)] = 3$$

(Remember, $\sum (h)$ is not the total dominance effect of the genes, but the balance between the increasing and decreasing effects of dominance.)

The total additive effect of the five genes, *AA*, *BB*, *CC*, *DD* and *EE* is equal to 5, that is, $\sum (d) = 5$.

For $[d]$ to be equal to $[h]$, $r_d = \dfrac{\sum (h)}{\sum (d)} = \dfrac{3}{5}$

In other words, for $\overline{P}_1$ to be equal to $\overline{F}_1$, the degree of association must be equal to or greater than $\frac{3}{5}$.

The genotype of the P_1 generation would be: $+ + \ \ + + \ \ + + \ \ + + \ \ - -$.

(Note that we are using plus and minus signs, because of the five genes could be in the increasing or decreasing state.)

11.8.3 Heterosis and plant and animal breeding

The objective of plant and animal breeding is to change permanently the mean phenotype of a population for characters of economic importance, for example, high yields in cereals, large numbers of eggs in poultry, high yields of milk in dairy cattle, and early ripening in cereals. Such an objective may be achieved by selecting from segregating populations those individuals whose phenotypes are nearest the objective, and using these as the parents for the next generation.

The process of selection may take any of several different forms, for example, mass selection where only the top few per cent are included in the mating population.

Another type of selection occurs with line-selection or line-breeding. Line-breeding consists of selecting the best families rather than the best individuals, collecting their progeny, allowing the progeny to grow and mate at random but in isolation, and then collecting their progeny.

The basic aim of any breeding process is to increase the frequency of the advantageous alleles within the gene pool, and to decrease the frequency of disadvantageous alleles. The actual rate of change in allele frequencies will be dependent on the intensity of selection.

Such methods as mass selection and line-breeding may take many years to come to fruition. The advantage of inbreeding, intercrossing and selection is that during the course of inbreeding, poor lines can be excluded from the programme, leaving only those individuals that show least inbreeding depression. Because inbreeding not only allows the recessive undesirable alleles to be quickly expressed, but also the recessive desirable alleles, the plant or animal breeder can assess his experimental material at an earlier stage.

As the expression of the disadvantageous alleles and their subsequent removal from the gene pool is very rapid, the frequency of the advantageous alleles may be increased at a faster rate than in a mass-selection or line-breeding programme. Consequently, the inbreeding and intercrossing procedure in the production of F_1 hybrids is a valuable tool for the plant or animal breeder.

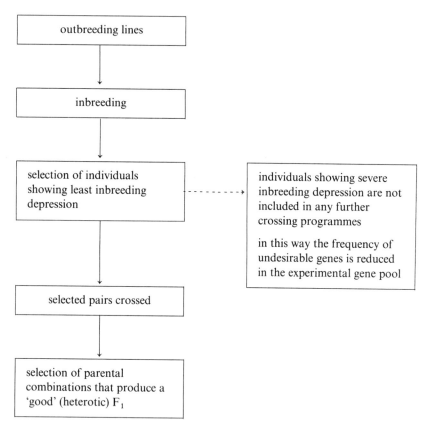

In the TV programme (Heterosis), we shall discuss the role of heterosis in the improvement of plants and animals further by talking to the breeders who use this and other methods.

11.9 Summary of Unit 11

We have considered the generation and description of continuous variation. Using mathematical models, the additive and dominance properties of genes may be described and their relative contributions to different generation means may be predicted from the Mendelian theory of inheritance and the laws of probability.

Of course, no model is a satisfactory description of any system unless its adequacy can be proven. In Section 11.5, we described the way in which the adequacy of an additive dominance model may be assessed.

We have said it before and we shall say it again, biometrical models have been built in order to explain continuous variation. Section 11.7 shows how an understanding of biometrical models can be applied to the solving of an important problem in plant and animal breeding.

Using an additive dominance model, hybrid vigour or heterosis can be explained in terms of additivity, dominance and gene association. From the information gained by using biometrical models, we can decide whether or not the observed heterosis can be 'fixed' in pure-breeding lines. The evidence usually suggests that it can.

The concepts and models that have been developed in Unit 11 are used again in Unit 12, so you are advised to remember these as far as possible.

Appendix 1   List of main symbols used in the text

A–a gene   a shorthand for a hypothetical locus, A, at which there are two alleles A and a.

m the mid-parent value. It is the natural mid-point from which to measure the deviation of the two homozygotes, the deviation being equal to the additive effects (d_a) of the gene. m is the phenotypic effect of all the common loci and reflects common environmental effects.

d the average additive effect of the gene over the range of environments in which the experimental material was reared; it is given by the deviation of either homozygous parent from the mid-parent value, m. By definition, the value of d is always positive.

h the dominance effect of the gene; it is equal to the deviation of the F_1 generation from the mid-parent value. Unlike the value of d, the value of h can be positive or negative according to the direction of dominance.

   Both d and h measure the average effects of the gene over all the environments in which the experimental material was reared.

$\overline{F}_1$, $\overline{P}_1$, $\overline{B}_1$, etc.   the mean of the F_1, P_1, B_1, etc. generations (spoken, P_1 bar).

$\sum (d)$ the sum total of all the individual additive effects of each allele: (d_+) denotes the sum of the increasing alleles; (d_-) denotes the sum of the decreasing alleles

$[d]$ the balance between the additive effects of the increasing and decreasing alleles $[d] = \sum (d_+) - \sum (d_-)$

$[h]$ the balance between the h effects in an increasing direction and the h effects in a decreasing direction $[h] = \sum (h_+) - \sum (h_-)$

$\sum (h)$ the sum of all the individual dominance effects of each allele: $\sum (h_+)$ denotes effects due to dominant increasing alleles; $\sum (h_-)$ denotes effects due to dominant decreasing alleles

r_d the degree of association of the genes

A-scaling test $= 2\overline{B}_1 - \overline{F}_1 - \overline{P}_1 = 0$, that is, the value of A will be zero within the limits set by sampling error

SE   standard error $= \sqrt{\left(\dfrac{s^2}{n}\right)}$

$\bar{x}$ estimate of the mean

μ expected value of the mean

s^2 sample variance

s sample standard deviation

N sample size

$V(x)$ variance of the mean

t Student's t-statistic

f frequency

Self-assessment questions

SAQ 1 In a particular variety of barley, the length of the straw is determined by the A–a gene. 10 individuals of the 2 parental lines and the F_1 heterozygote were grown in a randomized experimental design and the resulting straw lengths are listed below.

(a) Calculate the parameters m, d and h and the degree of dominance of the A–a gene.

(b) From your estimates of m, d and h, calculate the expected values of $\bar{F}_2$, $\bar{B}_1$ and $\bar{B}_2$.

P_1/cm	P_2/cm	F_1/cm
45.0	12.0	36.0
41.0	11.0	30.0
40.0	12.5	35.5
40.5	10.0	31.0
40.3	11.0	31.5
41.0	11.5	32.6
42.0	10.0	30.0
41.6	10.3	31.8
42.0	10.2	31.2
43.0	10.5	30.0

SAQ 2 Below we give data from three generations of barley. The character we wish to examine is the final weight of the seed or grain produced. Assuming that the three generations have been grown in a suitably replicated and randomized experimental design (a) calculate the parameters m, $[d]$ and $[h]$ and (b) calculate the potence ratio.

$$\bar{P}_1 = 5.405 \text{ g}$$
$$\bar{P}_2 = 2.050 \text{ g}$$
$$\bar{F}_1 = 2.395 \text{ g}$$

SAQ 3 Define the P_1, P_2, F_1, F_2, B_1 and B_2 generation means in terms of the parameters m, $[d]$ and $[h]$.

SAQ 4 Below we give data for four generations derived from a cross between two true-breeding lines of oats. The character we are examining is again the yield of grain. Assuming that the four generations have been grown in a suitably replicated and randomized experimental design:

(a) Calculate m, $[d]$ and $[h]$.

(b) Calculate the potence ratio.

(c) Is an additive dominant model an adequate description of the metrical relationship between genotypes?

	Sample number
$\bar{P}_1 = 58 \text{ g} \pm 0.185\,0$	20
$\bar{P}_2 = 26 \text{ g} \pm 0.140\,0$	20
$\bar{F}_1 = 50 \text{ g} \pm 0.180\,0$	20
$\bar{B}_1 = 55 \text{ g} \pm 0.570\,0$	20

SAQ 5 Using biometrical terms, explain the two possible causes of heterosis.

Answers to ITQs

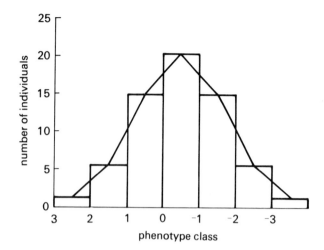

Figure 4

ITQ 2

$$m = \frac{KK + kk}{2} = \frac{48 + 26}{2} = 37$$

$$d_k = KK - m = 48 - 37 = 11$$

ITQ 3

$$m = \frac{BB + bb}{2} = \frac{84 + 16}{2} = 50$$

$$d_b = BB - m = 84 - 50 = 34$$

$$h_b = Bb - m = 60 - 50 = +10$$

ITQ 4 By definition, P_1 is always the larger of the two parents. Therefore,

$$\bar{P}_1 = 120 \quad \text{and} \quad \bar{P}_2 = 80$$

$$m = \frac{\bar{P}_1 + \bar{P}_2}{2} = \frac{120 + 80}{2} = 100$$

$$[d] = \bar{P}_1 - m = 120 - 100 = 20$$

ITQ 5

$$m = \frac{\bar{P}_1 + \bar{P}_2}{2} = \frac{15.0 + 10.0}{2} = 12.5$$

$$[d] = \bar{P}_1 - m = 15.0 - 12.5 = 2.5$$

$$[h] = \bar{F}_1 - m = 11.0 - 12.5 = -1.5$$

ITQ 6 The correct answer is (iii).

The potence ratio is a measure of the degree of dominance only when the degree of association, that is, r_d, is equal to 1, and all the h effects are unidirectional.

Let us see why this is.

On a one-gene model, then, $\dfrac{h}{d}$ is an accurate measure of the degree of dominance.

For differences in many genes, we can only measure $[h]$ and $[d]$ (where $[h]$ and $[d]$ are balance effects between increasing and decreasing alleles).

Now $[d] = r_d \sum (d)$, so unless r_d is equal to 1, $[d]$ is *not* a measure of the total additive effect of the genes.

Similarly, $[h] = \sum (h)$, but unless all the independent h effects have the same sign, $[h]$, that is, $\sum (h)$ is not the total dominance effect of the genes but represents the balance between the increasing and decreasing alleles.

When we cannot assume that these conditions are fulfilled, $\dfrac{[h]}{[d]}$ becomes a measure of the potence of the parents.

ITQ 7

(a)
$$m = \frac{\overline{P}_1 + \overline{P}_2}{2} = \frac{355 + 135}{2} = 245$$

$$[d] = \qquad 355 - 245 = 110$$

$$[h] = \qquad 300 - 245 = +55$$

The potence ratio $\dfrac{[h]}{[d]} = \dfrac{+55}{110} = +0.5$

(b) As $\overline{F}_2 = m + \frac{1}{2}[h]$, the mean of the F_2 generation would be expected to be $245 + 27.5 = 272.5$.

ITQ 8

(a)
$$m = \frac{9.028 + 6.318}{2} = 7.673$$

$$[d] = 9.028 - 7.673 = 1.355$$

$$[h] = 6.329 - 7.673 = -1.344$$

$$\frac{[h]}{[d]} = \frac{-1.344}{1.355} = -0.99$$

(b) In order to test the adequacy of an additive dominance model, we use an A-scaling test:

$$A = 2\overline{B}_1 - \overline{F}_1 - \overline{P}_1$$
$$A = 2(7.034) - (6.329) - (9.028) = -1.29$$
$$V(A) = 4V(\overline{B}_1) + V(\overline{F}_1) + V(\overline{P}_1)$$

As the variances given are s^2, it follows that the variance of the mean must be $\dfrac{s^2}{n}$.

Therefore
$$V(A) = 4(0.013\ 1) + (0.006\ 0) + (0.003\ 5) = 0.062\ 1$$

and the standard error of $A = \sqrt{V(A)} = \sqrt{0.062\ 1}$
$$= 0.25$$

Therefore, $\qquad A = -1.29 \pm 0.25$

and
$$t = \frac{1.29}{0.25} = 5.16$$

Reference to *STATS*, Table 3 on p. 57, shows quite clearly that a t value of 5.16 is significantly different from 0 at a level of probability of less than 0.001 [d.f. = 57], and therefore an additive dominance model is not an adequate description of the metrical relationship between genotypes.

ITQ 9

(a) The heterozygote would have the genotype: *Aa Bb Cc Dd Ee Ff*

(b)
$$[d] = \sum (d_+) - \sum (d_-) - \sum (d_-) = 4 - 2 = 2$$
$$[h] = \sum (h) = 3$$

Therefore, $\overline{F}_1$ will be greater than $\overline{P}_1$ because $[h]$ is greater than $[d]$.

(c) The farmer's own seeds would be an F_2 generation. The F_2 generation mean is less than the F_1 generation mean. Thus
$$\overline{F}_1 = m + [h]$$
$$\overline{F}_2 = m + \tfrac{1}{2}[h]$$

Consequently, his crop would be far less productive.

Second, because the F_2 generation undergoes segregation, the uniformity of the F_1 hybrid would be lost.

Even if the plant breeder did his utmost to keep the frequency of parental types in the hybrid seed as low as possible, it is inevitable with some systems that at least 2–5 per cent of the seeds would be parental types.

(d) If both parents occurred together, the F_1 hybrid could be re-created by other people. Consequently, seeds are collected from both parents separately, and care is taken to distribute the two sources of seed to different localities.

Answers to SAQs

SAQ 1 (*Objectives 2 and 3*)

(a)
$$\bar{P}_1 = 41.64 \text{ cm}$$
$$\bar{P}_2 = 10.90 \text{ cm}$$
$$\bar{F}_1 = 31.96 \text{ cm}$$
$$m = \frac{\bar{P}_1 + \bar{P}_2}{2} = \frac{41.64 + 10.90}{2} = 26.27$$
$$d_a = \bar{P}_1 - m = 41.64 - 26.27 = 15.37$$
$$h_a = \bar{F}_1 - m = 31.96 - 26.27 = +5.69$$
$$\frac{h_a}{d_a} = \frac{5.69}{15.37} = +0.37$$

(b)
$$\bar{F}_2 = m + \tfrac{1}{2}h_a = 29.12$$
$$\bar{B}_1 = m + \tfrac{1}{2}h_a + \tfrac{1}{2}d_a = 36.80$$
$$\bar{B}_2 = m + \tfrac{1}{2}h_a - \tfrac{1}{2}d_a = 21.44$$

SAQ 2 (*Objectives 4 and 5*)

(a)
$$m = \frac{\bar{P}_1 + \bar{P}_2}{2} = \frac{5.405 + 2.050}{2} = 3.728$$

$$[d] = \bar{P}_1 - m = 5.405 - 3.728 = 1.677$$
$$[h] = \bar{F}_1 - m = 2.395 - 3.728 = -1.333$$

(b)
$$\text{potence ratio} = \frac{[h]}{[d]} = \frac{-1.333}{1.677} = -0.795$$

SAQ 3 (*Objective 4*)

$$\bar{P}_1 = m + [d]$$
$$\bar{P}_2 = m - [d]$$
$$\bar{F}_1 = m + [h]$$
$$\bar{F}_2 = m + \tfrac{1}{2}[h]$$
$$\bar{B}_1 = m + \tfrac{1}{2}[d] + \tfrac{1}{2}[h]$$
$$\bar{B}_2 = m - \tfrac{1}{2}[d] + \tfrac{1}{2}[h]$$

SAQ 4 (*Objectives 4, 5 and 6*)

(a)
$$m = \frac{58 + 26}{2} = 42$$

$$[d] = 58 - 42 = 16$$
$$[h] = 50 - 42 = +8$$

(b)
$$\frac{[h]}{[d]} = \frac{8}{16} = +0.5$$

(c) To test the adequacy of the additive dominance model we apply an A-scaling test:

$$A = 2\bar{B}_1 - \bar{F}_1 - \bar{P}_1$$
$$A = 110 - 50 - 58 = +2$$

To test whether or not A is significantly different from zero, we need to find the standard error of A.

Standard error $= \sqrt{V(A)}$ where $V(A)$ is the variance of A

$$V(A) = 4V(\bar{B}_1) + V(\bar{F}_1) + V(\bar{P}_1)$$

Where $V(\bar{B}_1)$, $V(\bar{F}_1)$ and $V(\bar{P}_1)$ are the variances of the means of the B_1, F_1 and P_1 generations. Consequently,

$$4V(\bar{B}_1) = 4SE(B_1)^2 = 4(0.57)^2 = 1.299\,6$$
$$V(\bar{F}_1) = \ SE(F_1)^2 = \ (0.18)^2 = 0.032\,4$$
$$V(\bar{P}^1) = \ SE(P_1)^2 = \ (0.185) = 0.034\,2$$
$$V(A) = 1.366\,2$$
$$SE(A) = \sqrt{1.366\,2}$$
$$SE(A) = 1.169$$

$$t = \frac{2}{1.169} = 1.7$$

Consulting *STATS*, Table 3 on p. 57, shows that A is not significant at a level of probability of greater than 0.05, with 57 degrees of freedom.

An additive dominance model, therefore, is an adequate description of the metrical relationship between genotypes.

SAQ 5 (*Objective 7*) Heterosis is defined as the situation in which the F_1 generation mean is greater than the better of the two parents. How one decides which is the better parent depends on the particular character with which one is concerned. For example, if we consider the yield of grain, then heterosis will occur if $\bar{F}_1$ is larger than $\bar{P}_1$. If, on the other hand, we consider early ripening, then for heterosis to occur $\bar{F}_1$ must be smaller than $\bar{P}_2$. As

$$\bar{F}_1 = m + [h]$$
$$\bar{P}_1 = m + [d]$$
$$\bar{P}_2 = m - [d]$$

then, for heterosis to occur, $[h]$ must be positive or negative and greater than $[d]$.

Expressing this diagrammatically, for $\bar{F}_1$ to be greater than $\bar{P}_1$

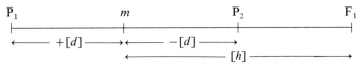

That is, $[h]$ is positive and greater than $[d]$.

For $\bar{F}_1$ to be smaller than $\bar{P}_2$

That is, $[h]$ is negative and greater than $[d]$.

To understand how $[h]$ may be greater than $[d]$, we need to look more closely at our definition of $[h]$ and $[d]$.

$[h] = \sum (h)$, where the values of h take a positive or negative sign. Consequently, unless *all* the hs are of equal sign, $\sum (h)$ is not the total dominance effect.

Similarly, $[d] = r_d \sum (d)$, where r_d is the degree of association of genes of like effect. Again, unless $r_d = 1$, then $[d]$ is not equal to the total additive effect. In this case, however, $\sum (d)$ is the *total* additive effect of the genes.

Let us now examine the two reasons why $\sum (h)$ could be greater than $\sum (d)$.

1 Overdominance If $\dfrac{h}{d}$ is greater than 1 at one or more loci, and dominance is preponderantly unidirectional then $\sum (h)$ could be greater than $\sum (d)$. Clearly, the extent to which $\sum (h)$ will be greater than $\sum (d)$ will be dependent on the value of each independent $\dfrac{h}{d}$ item.

2 Gene dispersion As $[d] = r_d \sum (d)$, unless all the genes of like effect are completely associated in one strain, that is, $r_d = 1$, $[d]$ will not be the total additive effect of the genes and [h] may appear to be the larger value, provided, of course, that the genes have preponderantly unidirectional dominance.

Where actual measurements of the total dominance and additive effects have been obtained, it has been shown that heterosis has been due to dispersed unidirectionally dominant genes in all instances.

12 Theories of Plant and Animal Breeding

S299
GENETICS

Contents

List of scientific terms used in Unit 12

Developed in this Unit	Page No.	Developed in this Unit	Page No.
additive genetic variance	561	gene pool	576
		heritability	558
alien gene pool	578	narrow-sense heritability	565
biparental progeny generation	569	parent–offspring covariance	573
broad-sense heritability	559	parent–offspring regression	574
dominance genetic variance	561		
		response to selection	567
environmental variance	559	selection differential	567

Objectives for Unit 12

After studying this Unit you should be able to:

1 Define, recognize the best definition of, and place in the correct context, the items in the list of scientific terms above.

2 Derive the variance of an F_2 generation for a single-gene difference in terms of d, h and E, and extend this derivation to include D, H and E (i.e. variance for differences in many genes).
(SAQ 1)

3 Derive the variance of the B_1 and B_2 generations for a single-gene difference.
(SAQ 2)

4 Demonstrate how D and H can be estimated by using the variances of F_2, B_1 and B_2 generations.
(SAQ 2)

5 Define broad-sense heritability in an F_2 generation in terms of the parameters, D, H and E.
(SAQ 1)

6 Define narrow-sense heritability in an F_2 generation in terms of the parameters D, H and E.
(SAQ 2)

7 Calculate the parameters, D, H, E, h_B^2 and h_N^2 from appropriate data.
(SAQs 1, 2 and 3)

8 Demonstrate the predictive properties of the narrow-sense heritability and the selection differential by calculating the response to selection.
(SAQ 3)

9 Define D_R and H_R for a random-mating population in terms of d, h, p and q. (Note: you are *not* expected to derive these expressions from first principles.)
(SAQ 4)

10 Define the broad-sense heritability in terms of D_R, H_R and E.
(SAQ 4)

11 Define the narrow-sense heritability in terms of D_R, H_R and E.
(SAQ 4)

12 Calculate the narrow-sense heritability in a random-mating population using the parent–offspring regression and the standard error of this estimate.
(SAQ 5)

13 Explain why the narrow-sense heritability changes over several generations of selection.

14 Discuss the objectives of plant and animal improvement and the major concepts involved, by referring to breeding programmes illustrated in TV programmes 'The Genetic Manipulation of Wheat', 'Heterosis' and 'Plant and Animal Breeding'.

Study guide for Unit 12

Unit 12 is very much an extension of Unit 11; it is, therefore, essential that you understand *all* the concepts that were defined and/or derived in Unit 11, *and* that you are able to derive such concepts from first principles. For example, you should be able to derive the means of F_2, B_1 and B_2 generations in terms of the parameters m, d and h.

In Unit 12 we describe continuous variation in terms of variances and examine the concepts involved in plant and animal improvement.

We realize that many people are intimidated by mathematics, but we can assure you that once you have mastered the algebra, which we work through slowly and logically, you will have achieved a far greater understanding of the genetic principles of plant and animal breeding. You will also, more than likely, feel a strong sense of satisfaction and achievement when you are able to say, 'Ah yes, the total genetic variance in an F_2 generation is $\frac{1}{2}D + \frac{1}{4}H$', and really know what this means. We do not want you to learn all the algebraic derivations but merely to understand their underlying genetic and statistical rationale.

The expressions you should be able to derive are:

1 The variance of an F_2 generation in terms of d, h, D, H and E. (Section 12.1.1)

2 The variance of B_1 and B_2 generations in terms of d, h, D, H and E. (Section 12.1.2)

3 The parent–offspring covariance. (Section 12.5.1)

4 The parent–offspring regression. (Section 12.5.2)

You will not be expected to derive any of the algebra in Sections 12.3.1 and 12.3.2. However, you will be expected to explain the significance of the quantities D_R and H_R, and to understand how these parameters take into account arbitrary gene frequencies. Finally, if you are really in trouble with the algebra, we suggest you ask your tutor for help.

TV programmes 'The Genetic Manipulation of Wheat', 'Heterosis' and 'Plant and Animal Breeding' are associated with the text; they deal with the practical problems of breeding plants and animals. These programmes are of particular importance as we have divided our treatment of plant and animal breeding into theory (text) and practice (TV programmes).

In the *Broadcast Notes** for TV programme 'Plant and Animal Breeding', you will find data relating to the two breeding programmes that are discussed. We have also included in the *Notes* a series of self-assessment questions (SAQs) that test Objectives 12, 13 and 14 of this Unit. If possible, you should work through these before listening to Radio programme 'Heritability and Variance' as we shall be discussing the results (i.e. the answers to the SAQs) in detail there.

The SAQs testing Objectives 1–12 are found at the end of the main text as usual. Attempt them at the end of Sections to which they refer, or leave them until you have completed your study of the Unit.

Finally, you are advised to have the statistics text† handy as we shall make direct reference to it.

---

* The *Broadcast Notes* are part of the supplementary material supplied to Open University Students as part of the Course.

† The Open University (1976) S299 STATS *Statistics for Genetics*, The Open University Press. This text is to be used in parallel with the Units of the Course. We refer to it by its code, *STATS*.

12.0   Introduction to Unit 12

Plant and animal breeding started some five or six thousand years ago when neolithic man first collected specific individuals from species of wild plants and animals, and domesticated them. Even before it was realized that characters tend to be inherited, the unconscious selection of certain traits is likely to have taken place. For example, highly fertile animals would be kept in the herd longer and, therefore, would contribute more offspring to the stock, and so on.

Eventually, man must have realized that offspring and their parents often resemble one another, and that sibs and other relatives show remarkable similarities. At this stage people probably attempted to discover effective breeding methods for retaining desirable characters within their stocks, but in the absence of any 'genetic' principles, and with the failure to recognize the correct function of pollination and fertilization, predictions of the results of matings were often made by observing such phenomena as the phases of the moon, the speed and direction of the wind and the position of animals during copulation.

After the rediscovery of Mendel's work, pollination, fertilization and the relationship between offspring and their parents began to make sense. In 1906, Bateson coined the term 'genetics' to describe the science of heredity, and since that time genetic principles have been applied to the ever-increasing demand for new and more productive varieties of crop plants and domestic animals.

In the last 50 years, plant and animal breeding has become both economically important, and a powerful applied science.

For example, it was found that the barley variety Proctor yielded 20 per cent more than its parents, var. Kenya and var. Plumage Archer. This increase represents some 0.05 tonnes extra yield per hectare, and with over $1\frac{1}{2}$ million hectares of land put down to barley in this country, this is a tremendous increase in productivity.

With over half the world's population starving or suffering from malnutrition, the creation of new, more productive, varieties of crops and animals is obviously of importance. Also, as the Third World countries develop their own agricultural systems, new varieties that can tolerate climatic conditions different from our own are needed, for example, drought-resistant varieties of grasses and cereals.

As present varieties of animals and crops succumb to diseases, new resistant strains must be produced. For example, in TV programme 'The Genetic Manipulation of Wheat', we saw how the yellow rust-resistant variety of wheat, Compair, has been produced by genetic manipulation.

Unit 12 is not a handbook of crop plant and animal husbandry; it is a rundown of the genetic principles of selection in plant and animal breeding. Neither are we trying to pretend that practical plant and animal breeders calculate all the parameters that we shall discuss in this Unit; however, breeders do rely on a basic understanding of these concepts in setting up breeding programmes that are often both costly and time consuming. As we mentioned in the Study guide, the practical aspects of plant and animal breeding are discussed in the three TV programmes.

To end this Introduction, we quote from a list of conclusions and recommendations from a seminar on agricultural genetics for Latin America, which was held in Maracay, Venezuela, in 1969:

> Institutes responsible for advanced agricultural education should intensify the teaching of genetics as applied to the improvement of plants and animals both at the undergraduate and at the postgraduate levels.

Unit 12 could well have been written to fulfil this objective!

12.1   Heritability

Before we can begin to understand the genetic principles of plant and animal breeding, we must consider the objectives and the genetic consequences of a selection programme. Plant and animal breeding depends on two major concepts, first, that differences do occur within the character in which we are interested, and second, that such differences are, at least in part, genetically determined.

heritability

558

Well how does one actually go about producing new varieties? In the process of domestication, many new strains of crop plants and domestic animals have been made simply by selecting variants within populations. For example, when sugar beet, *Beta maritima*, was first grown in this country about 150 years ago as a source of sucrose, the mean percentage of sucrose obtained was around 7 per cent. By selecting those individuals with much higher levels of sucrose as parents for the next generation, the mean was raised to 16 per cent by around 1910. Similarly, by selecting suitable variants of the wild čabbage, *Brassica oleracea*, plant breeders have been able to produce new and more useful varieties of crop plants. Thus, sprouts, broccoli, cabbage, cauliflower and curly kale, were all derived from variants of the wild cabbage.

Now, selecting among the two or, at the most, three alternative phenotypes as in the case of a difference in a single gene is a relatively easy task because discontinuities are very obvious. For example, consider the history of one particular variety of sweet pea. Before 1900, all known varieties of sweet pea had medium-sized flowers with plain upright petals. Around 1900, a mutation occurred in var. Prima Donna that caused the flowers to be large with curly upright petals. As the mutation was recessive, the change in floral morphology occurred only in the next and successive generations of selfing to the one in which the mutation originally arose. Let us denote the genotype of var. Prima Donna as *HH*, and that of the homozygous recessive mutant as *hh*. As *hh* is homozygous and recessive, it will breed true. The production of a new variety with *hh* characteristics is dependent on merely producing sufficient seeds for the horticultural market. Eventually, var. Countess Spencer was produced from this mutation.

But what if a character is determined by more than one gene and the distribution of phenotypic classes is continuous? Selection must be based on the phenotypes of the breeding parents or their relatives. But how accurately do phenotypic differences reflect genotypic differences?

The relative agreement between phenotypic differences and genotypic differences is given by the coefficient of broad-sense heritability, or more simply, the *broad-sense heritability*.

broad-sense heritability

> QUESTION From your knowledge of statistics and the description of continuous variation in Unit 11, Sections 11.1 and 11.2, can you suggest what parameters are used to define the broad-sense heritability?
>
> ANSWER The way in which continuous variation may be measured is by the mean and variance of its distribution. Thus, the observed variation may be measured as the phenotypic variance, which itself is composed of the variance due to genetic differences, that is, the genetic variance, and the variance due to environmental differences, that is, the environmental variance. The ratio
>
> $$\frac{V_g}{V_g + V_e}$$
>
> where V_g is the genetic variance and $V_g + V_e$ is the phenotypic variance, is a measure of the broad-sense heritability. Note that the phenotypic variance is composed of both the genetic variance, V_g, and the environmental variance, V_e.

environmental variance

The broad-sense heritability is itself denoted by h_B^2. You should note that the 2 does not mean that the value is squared, but that it is a variance component. So

$$h_B^2 = \frac{\text{genetic variance}}{\text{phenotypic variance}}$$

and this ratio tells us just how much of the phenotypic variation is genetically determined—clearly a most useful genetic concept. But how can we partition the observed or measured variation into its respective components?

> **ITQ 1** The genetic variance of straw length in an F_2 generation of oats is found to be 125 cm², and the environmental variance 25 cm². Calculate the broad-sense heritability, h_B^2.

The answers to the ITQs are on p. 581.

12.1.1 The broad-sense heritability in an F_2 generation

In Unit 11, continuous variation was described in terms of the additive (d) and dominance (h) effects of the genes. In this Section, we shall examine how these effects contribute to the variance of the measured character, and see how the total phenotypic variance may be partitioned into its components.

From Unit 11, you should realize that by using lines that are homozygous for different alleles of the same genes to initiate a breeding programme, the allele frequencies in all subsequent generations are set at a known value. In this Section, we shall consider the situation in which allele frequencies have been set at $p = q = 0.5$. Later, we shall extend these general principles to examples in which allele frequencies can take arbitrary values (Section 12.3.2).

The broad-sense heritability has been defined as the proportion of the total phenotypic variance that is genetic in origin. But how can we find the genetic and environmental variances?

You will recall from Unit 11, Section 11.3.5, that for a difference in a single gene the mean of an F_2 generation, $\overline{F}_2$, is given by $m + \frac{1}{2}h$. Let us see how this was obtained:

F_1		Aa	
F_2 genotypes	AA	Aa	aa
frequency	$\frac{1}{4}$	$\frac{1}{2}$	$\frac{1}{4}$
phenotype	$m + d$	$m + h$	$m - d$

AA contributes $\frac{1}{4}(m + d)$ to the F_2 generation mean,

Aa contributes $\frac{1}{2}(m + h)$ to the F_2 generation mean, and

aa contributes $\frac{1}{4}(m - d)$ to the F_2 generation mean.

Therefore, the F_2 mean, $\overline{F}_2$, is given by $m + \frac{1}{2}h$.

Now, to calculate the variance of an F_2 generation, we must derive the expressions for the sums of squares of deviations of the phenotypes ($m + d$, $m + h$ and $m - d$) from the F_2 generation mean, $m + \frac{1}{2}h$. As the mid-parent value is a constant, appearing in all four terms, we can ignore m in deriving the variances. To take an example: the phenotypic value of AA individuals is given by $m + d$ and the square of the deviation of $m + d$ from the F_2 mean, $m + \frac{1}{2}h$, is given by $[(m + d) - (m + \frac{1}{2}h)]^2$ $= (d - \frac{1}{2}h)^2$, the ms cancelling out. However, because only one-quarter of the F_2 generation are AA, their contribution to the F_2 variance is given by $\frac{1}{4}(d - \frac{1}{2}h)^2$. Similarly, the contribution of Aa to the F_2 variance is given by $\frac{1}{2}(h - \frac{1}{2}h)^2$ and the contribution of aa to the F_2 variance is given by $\frac{1}{4}(-d - \frac{1}{2}h)^2$.

If these three expressions are multiplied out

AA contributes $\frac{1}{4}(d - \frac{1}{2}h)^2 = \frac{1}{4}(d^2 + \frac{1}{4}h^2 - dh) = \frac{1}{4}d^2 + \frac{1}{16}h^2 - \frac{1}{4}dh$

Aa contributes $\frac{1}{2}(h - \frac{1}{2}h)^2 = \frac{1}{2}(\frac{1}{2}h)^2 = \frac{1}{2}(\frac{1}{4}h^2) = \frac{1}{8}h^2$

aa contributes $\frac{1}{4}(-d - \frac{1}{2}h)^2 = \frac{1}{4}(d^2 + \frac{1}{4}h^2 + dh) = \frac{1}{4}d^2 + \frac{1}{16}h^2 + \frac{1}{4}dh.$

By summing, we obtain

$\frac{1}{2}d^2 + \frac{1}{4}h^2$ (dh terms cancel one another out!)

Note: the dh term is the covariance obtained on the expansion of the binomial equation, $(p + q)^2$.

Therefore, the contribution of one gene to the variance in an F_2 generation is equal to $\frac{1}{2}d^2 + \frac{1}{4}h^2$.

If K genes determine a character, then the variance is the sum of all the independent variances of each gene. Thus:

$$\frac{1}{2}d_a^2 + \frac{1}{2}d_b^2 + \frac{1}{2}d_c^2 \cdots \frac{1}{2}d_k^2 = \frac{1}{2}\sum(d^2)$$

$$\frac{1}{4}h_a^2 + \frac{1}{4}h_b^2 + \frac{1}{4}h_c^2 \cdots \frac{1}{4}h_k^2 = \frac{1}{4}\sum(h^2)$$

For ease of presentation

$$\sum(d^2) \text{ is symbolized by } D;$$
$$\sum(h^2) \text{ is symbolized by } H.$$

Thus, the heritable or genetic variance in an F_2 generation is equal to $\frac{1}{2}D + \frac{1}{4}H$.

The heritable variance in the F_2 generation is, therefore, divisible into two components, the D component, which is the *additive genetic variance* because it depends on $\sum (d^2)$, and the H component, which is the *dominance genetic variance* because it depends on the $\sum (h^2)$ effects.

additive genetic variance
dominance genetic variance

At this stage we should point out that we have used a slightly different method of deriving the variance from the one you will be used to in *STATS*. The expression that we have used is

$$\sigma^2 = \sum_{i=1}^{k} f_i(x_i - \mu)^2$$

where f_i is the frequency of the ith variate,

μ is the true mean of the population,

$x_i - \mu$ is the deviation of the ith variate from the true mean μ, and

σ^2 is the population variance.

For example, the mean of AA individuals is $m + d$, the F_2 generation mean is $m + \frac{1}{2}h$, and the frequency of AA individuals in an F_2 generation is one-quarter. Thus, where AA individuals are the ith variate,

$$f_i(x_i - \mu)^2 = \frac{1}{4}(d - \frac{1}{2}h)^2$$

the ms cancelling out.

You should be aware that we are using proportionate frequencies fixed by the Mendelian theory of inheritance, which sum to unity. In these circumstances the number of observations is equal to the number of degrees of freedom, μ being the true mean of the population. As these conditions are met only in models, you should not try to use this expression for any calculation of variance that is not purely theoretical.

The total phenotypic variance must also contain an environmental component, V_e, which we now denote by E. No two individuals within any one generation will be exposed to exactly the same environment, and this leads to differences that are environmental in origin. The phenotypic variance of an F_2 generation is, therefore, given by

$$V(F_2) = \frac{1}{2}D + \frac{1}{4}H + E$$

QUESTION In an experiment, plants of the two pure-breeding parents P_1 and P_2, their F_1 hybrid generation and the F_2 generation are all grown together in a randomized experimental design. From the variances of the four generations it is possible to partition the F_2 variance into its genetic component and its environmental component. Can you see why we are able to estimate the environmental variance from the variances of the P_1, P_2 or F_1 generations?

ANSWER As the three generations consist of non-segregating lines, any observed variation must be caused by environmental differences only. Hence, the environmental variance may be estimated because individual members of each generation have been grown together in a randomized experimental design.

QUESTION In the previous Section, Section 12.1, we defined the broad-sense heritability, h_B^2, as being that proportion of the total phenotypic variance that is genetically determined. Define the broad-sense heritability in terms of the parameters D, H and E.

ANSWER

$$h_B^2 = \frac{\text{heritable or genetic variance of the } F_2}{\text{total phenotypic variance of the } F_2}$$

Therefore, the broad-sense heritability, h_B^2, for an F_2 generation

$$= \frac{\frac{1}{2}D + \frac{1}{4}H}{\frac{1}{2}D + \frac{1}{4}H + E}$$

Let us work through a practical example. We give opposite the seed weights for three generations of rye, which have been grown together in a randomized experimental design. Using this data, we can now calculate the broad-sense heritability, h_B^2.

generation	variance
P_1	$= 5.80 \text{ g}^2$
P_2	$= 5.60 \text{ g}^2$
F_2	$= 35.00 \text{ g}^2$

We know that the phenotypic variance of the F_2 generation is equal to $\frac{1}{2}D + \frac{1}{4}H + E$, and this equals 35.00.

We also know that the phenotypic variance of the P_1 and P_2 generations is equal to E. Taking the mean of the two parental generations as an estimate of E, that is, 5.70, $\frac{1}{2}D + \frac{1}{4}H$ is equal to $35.00 - 5.70$ which equals 29.30. Thus

$$\frac{1}{2}D + \frac{1}{4}H \text{ is equal to } 29.30$$

and

$$\frac{1}{2}D + \frac{1}{4}H + E \text{ is equal to } 35.00.$$

As h_B^2 is equal to

$$\frac{\frac{1}{2}D + \frac{1}{4}H}{\frac{1}{2}D + \frac{1}{4}H + E}$$

the broad-sense heritability is equal to

$$\frac{29.30}{35.00} = 0.84$$

that is,

$$h_B^2 = 0.84$$

In other words, 84 per cent of the phenotypic variation is genetic in origin.

> **ITQ 2** We give below data for straw length in a variety of wheat. Three generations, P_1, P_2 and F_2, have been grown in a randomized experimental design.
>
> | generation | variance |
> |---|---|
> | P_1 | $= 20 \text{ cm}^2$ |
> | P_2 | $= 10 \text{ cm}^2$ |
> | F_2 | $= 35 \text{ cm}^2$ |
>
> Calculate the broad-sense heritability. What does the value of this estimate tell you about the variation in straw length?

You should now attempt SAQ 1 (p. 580).

12.1.2 The narrow-sense heritability in an F_2 generation

In the Introduction, we stated that offspring resembled their parents and that sibs tended to show remarkable similarities. But what are these similarities?

If we breed another generation by taking parents at random from an F_2 generation, then the mean of the progeny generation, and the genetic variance, will be the same as for an F_2 generation, that is, $m + \frac{1}{2}[h]$ and $\frac{1}{2}D + \frac{1}{4}H$. This is because both generations have equal allele frequencies and are in Hardy–Weinberg equilibrium (see Units 9 and 10, Section 9.4.2).

However, it is genes that are inherited not genotypes, so, at meiosis, the dominance relationships in the parental genotypes are destroyed and recreated at random within the progeny generation. In other words, the dominance effects are redistributed at random in the progeny generation.

	AA $\frac{1}{4}$	Aa $\frac{1}{2}$	aa $\frac{1}{4}$	
AA $\frac{1}{4}$	P_1 $m + d$ $\frac{1}{16}$	B_1 $m + \frac{1}{2}d + \frac{1}{2}h$ $\frac{1}{8}$	F_1 $m + h$ $\frac{1}{16}$	family type mean frequency
Aa $\frac{1}{2}$	B_1 $m + \frac{1}{2}d + \frac{1}{2}h$ $\frac{1}{8}$	F_2 $m + \frac{1}{2}h$ $\frac{1}{4}$	B_2 $m - \frac{1}{2}d + \frac{1}{2}h$ $\frac{1}{8}$	family type mean frequency
aa $\frac{1}{4}$	F_1 $m + h$ $\frac{1}{16}$	B_2 $m - \frac{1}{2}d + \frac{1}{2}h$ $\frac{1}{8}$	P_2 $m - d$ $\frac{1}{16}$	family type mean frequency

Assuming random mating, the frequency of $aa \times aa$ crosses is given by the product of their relative frequencies in the F_2 generation. Thus, the frequency of $aa \times aa$ crosses is given by $\frac{1}{4} \times \frac{1}{4} = \frac{1}{16}$.

As $aa \times aa$ crosses are of P_2 family type, we can say that $\frac{1}{16}$ of the progeny will be of P_2 family type, with a mean of $m - d$. Similarly, the frequency of $aa \times AA$ crosses is equal to $\frac{1}{4} \times \frac{1}{4} = \frac{1}{16}$, and $\frac{1}{16}$ of its progeny will be of $AA \times aa$ family type.

QUESTION What is the family type of $AA \times aa$ crosses?

ANSWER $AA \times aa$ crosses give rise to the F_1 family type, with a mean of $m + h$.

Now, if we sum all the relative contributions of the different family types, we find that the total variance is again equal to $\frac{1}{2}D + \frac{1}{4}H$ and that the mean is again equal to $m + \frac{1}{2}[h]$; but the dominance effects of the genes have been redistributed.

For example, take $AA \times aa$ crosses:

Mean of the cross in terms of

	m	d	h

$$
\begin{array}{ccc}
AA & \times & aa \\
m+d & & m-d
\end{array}
$$
m — —

$\downarrow$

$$
\begin{array}{c}
Aa \\
m+h
\end{array}
$$
m — h

Again, consider $Aa \times Aa$ crosses:

$$
\begin{array}{ccc}
Aa & \times & Aa \\
m+h & & m+h
\end{array}
$$
m — h

$$
\begin{array}{ccc}
\frac{1}{4}AA & \frac{1}{2}Aa & \frac{1}{4}aa \\
\frac{1}{4}(m+d) & \frac{1}{2}(m+h) & \frac{1}{4}(m-d)
\end{array}
$$
m — $\frac{1}{2}h$

Similarly, consider $AA \times AA$ crosses:

$$
\begin{array}{ccc}
AA & \times & AA \\
m+d & & m+d
\end{array}
$$
m d —

$\downarrow$

$$
\begin{array}{c}
AA \\
m+d
\end{array}
$$
m d —

So, if we look at individuals within the F_2 generation and at their families in the progeny generation, then we find that because of the redistribution of the dominance effects of the genes, these effects cannot cause any systematic resemblances between the parents and the offspring in the two generations. Only the additive effects of the genes, that is, the relative frequency of increasing and decreasing alleles, are expected to reappear in the progeny in such a way as to cause similarities between the parents and their offspring in the two generations.

Now, when a plant or animal breeder selects individuals from an F_2 generation as potential parents for the next generation, he will not do this at random. Far from it. He will select those individuals that are closest to his objective or desired result.

Figure 1 Distribution of heights of (a) initial and (b) selected populations of a variety of barley.

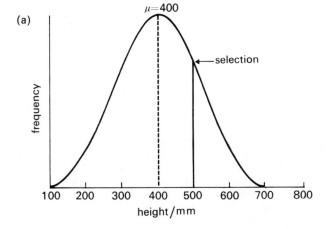

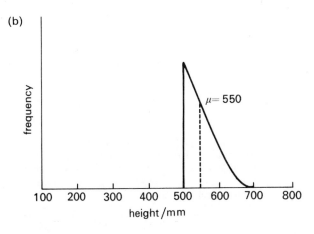

For example, let us say that the F_2 genotypic distribution of height in a variety of barley has a mean of 400 mm. (We are using genotypic distribution to eliminate environmental effects at this stage.) The plant breeder selects those individuals that are taller than 500 mm, the mean of the distribution being 550 mm (see Fig. 1). Now, the breeder wants to know what the mean of the progeny generation will be. (Remember, we have selected specific individuals and have not taken plants at random.) Because of the redistribution of the dominance genetic effects, only the additive effects of the genes will cause any similarities between the selected group of parents and their progeny.

For example, if all the variation is additive genetic in origin, there will be perfect correspondence between the distributions of parents and progeny (see Fig. 2).

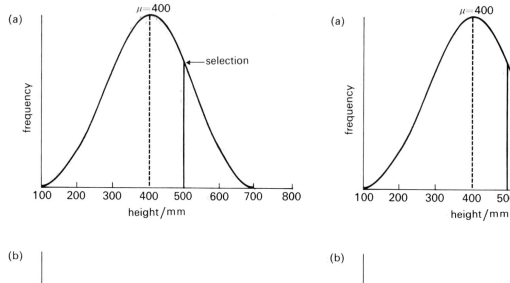

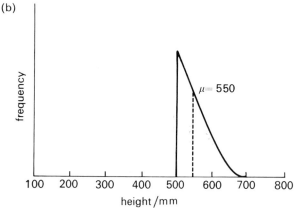

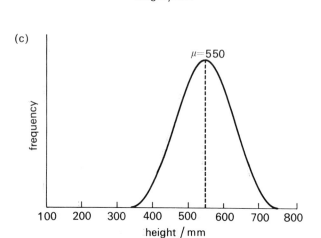

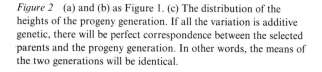

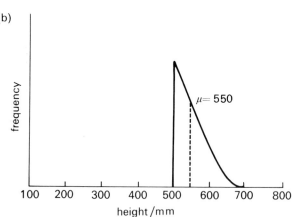

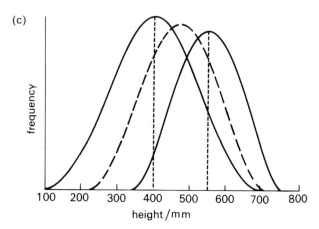

Figure 2 (a) and (b) as Figure 1. (c) The distribution of the heights of the progeny generation. If all the variation is additive genetic, there will be perfect correspondence between the selected parents and the progeny generation. In other words, the means of the two generations will be identical.

Figure 3 (a) and (b) as Figure 1. (c) The distribution of the heights of the progeny generation. If only a proportion of the variation is additive genetic, then the mean and the distribution of the progeny will lie somewhere between the mean and distribution of the initial population and the mean and distribution of the selected population.

If, on the other hand, only part of the variation is additive genetic, then the mean and the distribution of the character in the progeny generation will lie somewhere between the initial population and the selected group of parents (see Fig. 3).

So, in order to be able to predict any advance under selection, the breeder must be able to calculate the additive genetic variance.

Now, if we express the additive genetic variance, $\frac{1}{2}D$, as a proportion of the total phenotypic variance, this relationship is termed the *narrow-sense heritability*, h_N^2.

narrow-sense heritability

QUESTION Define the narrow-sense heritability, h_N^2, for an F_2 generation, in terms of the parameters D, H and E.

ANSWER

$$h_N^2 = \frac{\text{additive genetic variance}}{\text{phenotypic variance}}$$

$$h_N^2 = \frac{\frac{1}{2}D}{\frac{1}{2}D + \frac{1}{4}H + E}$$

The narrow-sense heritability tells us how much, or what proportion of the phenotypic differences are additive genetic in origin and will, therefore, reappear in the progeny generation in such a way as to cause similarities between the parental and offspring generations—the parameter that plant and animal breeders require to evaluate their breeding material.

If we now examine the components of the variance in backcross generations, we shall see how D and H and the narrow-sense heritability may be estimated in an F_2 generation.

In backcrossing, the B_1 and the B_2 progeny will be half homozygous and half heterozygous for each allele pair by which the parental lines differ (see Unit 11, Section 11.3.5).

QUESTION Using the principles shown in Section 12.1.1, derive the variances of (a) the B_1 and (b) the B_2 generation for a single-gene difference in terms of m, d and h.

ANSWER

(a) B_1 generation $\qquad AA \times Aa$

B_1 genotypes	AA	Aa
frequency	$\frac{1}{2}$	$\frac{1}{2}$
phenotype	$m + d$	$m + h$ $\qquad \overline{B}_1 = m + \frac{1}{2}d + \frac{1}{2}h$

AA contributes $\quad [(m + d) - (m + \frac{1}{2}d + \frac{1}{2}h)]^2 = (d - \frac{1}{2}d - \frac{1}{2}h)^2 = (\frac{1}{2}d - \frac{1}{2}h)^2$
$$= \tfrac{1}{4}d^2 + \tfrac{1}{4}h^2 - \tfrac{1}{2}dh$$

As only half of the B_1 generation are of AA type, their contribution to the B_1 variance is equal to

$$\tfrac{1}{2}(\tfrac{1}{4}d^2 + \tfrac{1}{4}h^2 - \tfrac{1}{2}dh) = \tfrac{1}{8}d^2 + \tfrac{1}{8}h^2 - \tfrac{1}{4}dh$$

Similarly, Aa contributes

$$[(m + h) - (m + \tfrac{1}{2}d + \tfrac{1}{2}h)]^2 = (\tfrac{1}{2}h - \tfrac{1}{2}d)^2$$
$$= \tfrac{1}{4}d^2 + \tfrac{1}{4}h^2 - \tfrac{1}{2}dh$$

Only half of the B_1 generation are of Aa type, and their contribution to the B_1 variance is equal to

$$\tfrac{1}{2}(\tfrac{1}{4}d^2 + \tfrac{1}{4}h^2 - \tfrac{1}{2}dh) = \tfrac{1}{8}d^2 + \tfrac{1}{8}h^2 - \tfrac{1}{4}dh$$

and the total variance of the B_1 generation is given by

$$V(B_1) = \tfrac{1}{4}d^2 + \tfrac{1}{4}h^2 - \tfrac{1}{2}dh = \tfrac{1}{4}(d - h)^2$$

(b) B_2 generation $\qquad aa \times Aa$

B_2 genotypes	aa	Aa
frequency	$\frac{1}{2}$	$\frac{1}{2}$
phenotype	$m - d$	$m + h$ $\qquad \overline{B}_2 = m - \frac{1}{2}d + \frac{1}{2}h$

aa contributes $\quad [(m - d) - (m - \tfrac{1}{2}d + \tfrac{1}{2}h)]^2 = (-\tfrac{1}{2}d - \tfrac{1}{2}h)^2$
$$= \tfrac{1}{4}d^2 + \tfrac{1}{4}h^2 + \tfrac{1}{2}dh$$

Only half of the B_2 generation are of aa type, and their contribution to the B_2 variance is equal to

$$\tfrac{1}{2}(\tfrac{1}{4}d^2 + \tfrac{1}{4}h^2 + \tfrac{1}{2}dh) = \tfrac{1}{8}d^2 + \tfrac{1}{8}h^2 + \tfrac{1}{4}dh$$

Similarly, Aa contributes

$$[(m + h) - (m - \tfrac{1}{2}d + \tfrac{1}{2}h)]^2 = (\tfrac{1}{2}h + \tfrac{1}{2}d)^2 = \tfrac{1}{4}d^2 + \tfrac{1}{4}h^2 + \tfrac{1}{2}dh$$

Only half of the B_2 generation are of Aa type, and their contribution to the B_2 variance is equal to

$$\tfrac{1}{2}(\tfrac{1}{4}d^2 + \tfrac{1}{4}h^2 + \tfrac{1}{2}dh) = \tfrac{1}{8}d^2 + \tfrac{1}{8}h^2 + \tfrac{1}{4}dh$$

The total variance of the B_2 generation is thus given by

$$V(B_2) = \tfrac{1}{4}d^2 + \tfrac{1}{4}h^2 + \tfrac{1}{2}dh = \tfrac{1}{4}(d + h)^2$$

Thus, the contribution of a single gene to the variance of the B_1 generation is given by $\tfrac{1}{4}d^2 + \tfrac{1}{4}h^2 - \tfrac{1}{2}dh$, and the variance of the B_2 generation is given by $\tfrac{1}{4}d^2 + \tfrac{1}{4}h^2 + \tfrac{1}{2}dh$.

The contributions of d and h are not separable in this form because they are covariant. If, however, the two backcross generations are added together, $V(B_1) + V(B_2)$, the contribution of the A–a locus is equal to

$$\tfrac{1}{4}d^2 + \tfrac{1}{4}h^2 - \tfrac{1}{2}dh + \tfrac{1}{4}d^2 + \tfrac{1}{4}h^2 + \tfrac{1}{2}dh = \tfrac{1}{2}d^2 + \tfrac{1}{2}h^2$$

If all the genes are independent in their transmission, that is, there is no linkage, then for differences in many genes, the heritable variance of the two backcross generations will be equal to $\tfrac{1}{2}\sum (d^2) + \tfrac{1}{2}\sum (h^2)$, which in terms of D and H, is equal to $\tfrac{1}{2}D + \tfrac{1}{2}H$.

You should have noticed that the difference between the heritable components of the F_2 variance, $\tfrac{1}{2}D + \tfrac{1}{4}H$, and the sum of the two backcross generations, $\tfrac{1}{2}D + \tfrac{1}{2}H$, is equal to $\tfrac{1}{4}H$. Once we know the variance of the F_2 generation and the two backcross generations, therefore, D, H and h_N^2 may be estimated.

Let us consider an example. The following data are the variances for the distributions of seed produced by various generations of barley derived from an initial cross between two pure-breeding lines. Let us see how D, H and h_N^2 can be estimated from these data.

$$V(F_2) = 130.5\ g^2$$
$$V(B_1) = 85.5\ g^2$$
$$V(B_2) = 98.5\ g^2$$
$$E = 42.0\ g^2$$

$$V(F_2) = \tfrac{1}{2}D + \tfrac{1}{4}H + E = 130.5$$
$$\tfrac{1}{2}D + \tfrac{1}{4}H = 130.5 - 42.0 = 88.5$$

QUESTION Calculate h_B^2.

ANSWER

$$h_B^2 = \frac{\tfrac{1}{2}D + \tfrac{1}{4}H}{\tfrac{1}{2}D + \tfrac{1}{4}H + E} = \frac{88.5}{130.5} = 0.68$$

$$V(B_1) + V(B_2) = \tfrac{1}{2}D + \tfrac{1}{2}H + 2E$$

(Note: there are two E components because each of the two generations is subject to an environmental variance.)

$$\tfrac{1}{2}D + \tfrac{1}{2}H + 2E = 85.5 + 98.5 = 184.0$$
$$\tfrac{1}{2}D + \tfrac{1}{2}H = 184.0 - 84 = 100.0$$

QUESTION What are the values of D and H and the narrow-sense heritability, h_N^2 ?

ANSWER
As

$$\tfrac{1}{2}D + \tfrac{1}{2}H = 100.0$$

and

$$\tfrac{1}{2}D + \tfrac{1}{4}H = 88.5$$
$$\tfrac{1}{4}H = 11.5$$

and

$$H = 46.0$$

As

$$\tfrac{1}{2}D + \tfrac{1}{2}H = 100.0$$

566

and

$$\tfrac{1}{2}H = 23.0$$
$$\tfrac{1}{2}D = 77.0$$

and

$$D = 154.0$$

The dominance genetic variance, H, is equal to 46, and the additive genetic variance, D, is equal to 154.

$$h_N^2 = \frac{\tfrac{1}{2}D}{\tfrac{1}{2}D + \tfrac{1}{4}H + E} = \frac{77}{130.5} = 0.59$$

Thus, 59 per cent of the variation in the F_2 generation is additive genetic in origin.

ITQ 3 Data of the variance for straw length of various generations of oats are given below. Calculate D, H and the narrow-sense heritability, h_N^2.

$$V(P_1) = 300 \text{ cm}^2$$
$$V(F_2) = 500 \text{ cm}^2$$
$$V(B_1) = 450 \text{ cm}^2$$
$$V(B_2) = 400 \text{ cm}^2$$

Before we consider other methods by which narrow-sense heritability may be estimated, let us first look at the predictive role of this parameter.

You should now attempt SAQ 2 (p. 580).

12.2 Heritability, the selection differential and the response to selection

To understand the importance of narrow-sense heritability, we need to examine briefly the methods used in plant and animal breeding. The important point is that superior individuals are selected as parents for the next generation. By superior we may mean larger or smaller plants, cows with higher or lower butter-fat content in milk or pigs with larger or smaller amounts of back fat. To generalize, however, plant and animal breeders select as their parental stock individuals that do not characterize the mean phenotype. In this way, the means of the different populations are directed towards either extreme of the phenotypic distribution of the characters in the population.

For example, let us suppose that the mean yield of milk in a herd of dairy cattle is 15 litres per day. The animal breeder hopes to produce a new, higher yielding variety. He selects those cows that produce 18 or more litres per day, the mean of the distribution being 20 litres, and uses these as his parental generation. Now, the measure of selection, or the *selection differential*, S, is the average superiority of the selected parents expressed as a deviation from the whole of the population from which they were selected.

selection differential

QUESTION What is the selection differential for milk yield in the dairy herd we have just quoted?

ANSWER The selection differential is the difference between the mean of the selected group of parents, and the mean of the whole parental generation before selection was applied. Thus, $S = 20 - 15$, and the selection differential is equal to 5 litres per day.

What we are really interested in, is predicting the mean yield of milk from the progeny generation derived from the selected group of parents; we call this the *response to selection*.

response to selection

The response to selection, R, may be defined as the difference between the progeny of the selected parents and the mean of the parental generation before selection was applied.

We know that if selection is to be of any use at all, then the variation in the selected group of parents must be, at least in part, additive genetic in origin. In other words, the response to selection will be directly proportional to the narrow-sense heritability. Provided that all the selected parents are equally represented in the progeny generation, then the response to selection, R, is equal to the product of the narrow-sense heritability, h_N^2, and the selection differential, S_d. That is,

$$R = h_N^2 \times S_d$$

For example, if we use the same data as for the previous QUESTION and ANSWER, and assume that $h_N^2 = 0.5$, then the response to selection is equal to $R = h_N^2 \times S_d = 0.5 \times 5 = 2.5$. In other words, the mean of the progeny generation will be 2.5 litres in excess of the mean of the parental generation before selection was applied. Thus, the mean of the progeny generation is given by, $15 + 2.5 = 17.5$ litres per day.

QUESTION Suppose the phenotypic distribution of grain yield in a variety of barley has a mean value of 10 g per plant. Individuals are selected that yield 12 g and over, the distribution having a mean of 14 g. Calculate the selection differential.

ANSWER $S_d = 14 - 10 = 4$ g. Thus, the selection differential is equal to 4 g.

QUESTION Let us now suppose that the additive genetic variance has been found to be 25 g^2, and the phenotypic variance 30 g^2. Calculate the narrow-sense heritability, h_N^2.

ANSWER

$$h_N^2 = \frac{\text{additive genetic variance}}{\text{phenotypic variance}} = \frac{25}{30} = 0.83$$

QUESTION Calculate the response to selection, R, in the above example.

ANSWER

$$R = h_N^2 \times S_d = 0.83 \times 4 = 3.32$$

QUESTION What is the new mean of the progeny generation?

ANSWER The response to selection is the deviation of the mean of the progeny generation from the mean of the parental generation before selection was applied. As the mean of the parental generation $= 10$ g and the value of $R = 3.32$ g, then the mean of the progeny generation will be $10 + 3.32 = 13.32$ g.

ITQ 4 Assuming that the selection differential and the narrow-sense heritability remain constant, how many generations of selection would it require to increase the mean from 10 to 19 g?

You probably calculated the answer to be 3 generations because the mean would increase by 3.32 g per generation of selection. However, in making this calculation we have assumed that the selection differential and the heritability remain constant throughout—assumptions that are not necessarily correct because both parameters may change as selection proceeds.

That the selection differential will decrease as selection proceeds is fairly easy to appreciate. For example, if selection is for high-yielding varieties of cereal, then by the sequential selection of higher ranking lines for the parental material, the variation within the parental material must also decrease. As the variation decreases so the selection differential will decrease. The selection differential must, therefore, be calculated from observations made at the time of the selection.

To understand the significance of the change in value of the narrow-sense heritability, we must analyse the quantities D and H once again.

You should now attempt SAQ 3 (p. 580).

12.3 The stability of the narrow-sense heritability

You should recall from Section 12.1.1, that the contribution of D and H to the phenotypic variance of an F_2 generation applied only when the allele frequencies are equal. Allele frequencies were set at $p = q = 0.5$, by using parental lines that were homozygous for different alleles of the same genes. In this way, the frequency of genotypes and their contribution to the variance in successive generations could be predicted according to Mendelian theory.

But the genotypes we have to work with in a breeding programme may not have equal allele frequencies. Indeed, the effects of directional selection* are to alter the initial allele frequency to some new level. What effect will this change in allele frequency have on our estimates of variance, additive genetic variance and, therefore, narrow-sense heritability?

12.3.1 Random-mating population—equal allele frequencies

Let us consider an F_2 generation in which the genotypes AA, Aa and aa occur with frequencies of $\frac{1}{4}, \frac{1}{2}$ and $\frac{1}{4}$. If we produce another generation by taking pairs of parents at random from the F_2 generation, the total heritable variance is again equal to $\frac{1}{2}D + \frac{1}{4}H$. This follows because an F_2 generation is one in which random mating occurs; the allele frequencies are equal and the population is in Hardy–Weinberg equilibrium (see Section 12.1.2). When random mating takes place in an F_2 generation, there are no changes in allele or genotype frequencies and, hence, there is no change in the mean or the variance.

Such a generation is termed a *biparental progeny generation* or a BIPs generation. If we breed yet another generation from a BIPs generation, taking pairs of parents at random from the first BIPs generation, the genetic variance is again $\frac{1}{2}D + \frac{1}{4}H$. $\frac{1}{2}D + \frac{1}{4}H$ is, therefore, characteristic of a random-mating population where $p = q = 0.5$.

biparental progeny generation

Let us take a closer look at a BIPs generation.

	AA $\frac{1}{4}$	Aa $\frac{1}{2}$	aa $\frac{1}{4}$	
AA $\frac{1}{4}$	P_1 $m + d$ $\frac{1}{16}$	B_1 $m + \frac{1}{2}d + \frac{1}{2}h$ $\frac{1}{8}$	F_1 $m + h$ $\frac{1}{16}$	family mean frequency
Aa $\frac{1}{2}$	B_1 $m + \frac{1}{2}d + \frac{1}{2}h$ $\frac{1}{8}$	F_2 $m + \frac{1}{2}h$ $\frac{1}{4}$	B_2 $m - \frac{1}{2}d + \frac{1}{2}h$ $\frac{1}{8}$	family mean frequency
aa $\frac{1}{4}$	F_1 $m + h$ $\frac{1}{16}$	B_2 $m - \frac{1}{2}d + \frac{1}{2}h$ $\frac{1}{8}$	P_2 $m - d$ $\frac{1}{16}$	family mean frequency

You will recall from Section 12.1.2 how this table is built up, but we shall run through it again here. An F_2 generation consists of $\frac{1}{4}AA$, $\frac{1}{2}Aa$ and $\frac{1}{4}aa$ individuals. The frequency of $AA \times AA$ crosses, assuming random mating, is given by the product of their relative frequencies in the F_2 generation. Thus, the frequency of $AA \times AA$ crosses is $\frac{1}{4} \times \frac{1}{4} = \frac{1}{16}$ and $\frac{1}{16}$ of the progeny families will consist of AA individuals.

As AA individuals are known as P_1, we can say that $\frac{1}{16}$ of the BIPs generation will be of P_1 family type with a mean of $m + d$.

Similarly, the frequency of $AA \times Aa$ crosses is given by the product of their relative frequencies in the F_2 generation. Thus, the frequency of $AA \times Aa$ crosses is $\frac{1}{4} \times \frac{1}{2} = \frac{1}{8}$; therefore, $\frac{1}{8}$ of the BIPs generation are families of the type $AA \times Aa$.

QUESTION What is the family type of $AA \times Aa$ crosses?

ANSWER $AA \times Aa$ crosses give rise to the B_1 type of family.

* Directional selection occurs when selection favours some phenotype that does not characterize the mean of the population.

QUESTION What is the mean of a B_1 generation?

ANSWER The mean of the B_1 generation is $m + \frac{1}{2}d + \frac{1}{2}h$

(If you had difficulty with this QUESTION and ANSWER refer back to Unit 11, Section 11.3.5.)

F_2 generations, and BIPs derived from F_2 generations, have identical means and variances; both have the same allele and genotype frequencies because both are in Hardy–Weinberg equilibrium.

12.3.2 Random-mating populations—arbitrary allele frequencies

Now, let us extend the general principles that we discussed in Unit 11, Section 11.3.2, to the situation in which the allele frequencies are not equal and, of course, the genotype frequencies are not those characteristic of an F_2 generation. Consider, once again, the A–a gene. Provided that the population is in Hardy–Weinberg equilibrium, the three genotypes will occur with frequencies of p^2, $2pq$ and q^2, where the frequency of the a allele is given by q, and the frequency of the A allele is given by p. This means that we can take our original table of the BIPs generation, where the frequencies of AA, Aa and aa were $\frac{1}{4}$, $\frac{1}{2}$ and $\frac{1}{4}$, and replace these frequencies with p^2, $2pq$ and q^2.

random-mating populations

Let us see the effect of this substitution*:

		AA p^2	Aa $2pq$	aa q^2	
AA		P_1	B_1	F_1	family
		d	$\frac{1}{2}(d + h)$	h	mean
p^2		p^4	$2p^3q$	p^2q^2	frequency
Aa		B_1	F_2	B_2	family
		$\frac{1}{2}(d + h)$	$\frac{1}{2}h$	$\frac{1}{2}(-d + h)$	mean
$2pq$		$2p^3q$	$4p^2q^2$	$2pq^3$	frequency
aa		F_1	B_2	P_2	family
		h	$\frac{1}{2}(-d + h)$	$-d$	mean
q^2		p^2q^2	$2pq^3$	q^4	frequency

For a BIPs generation with equal allele frequencies, we said that the contribution of P_1 to the generation mean would be $\frac{1}{16}d$. In this case, the contribution of the P_1 family to the population mean would be p^4d; the contribution of the F_1 family will be p^2q^2h. The full set of contributions are given by

$$P_1 = p^4d$$
$$B_1 = 2 \times 2p^3q(\tfrac{1}{2}d + \tfrac{1}{2}h)$$
$$F_1 = 2 \times p^2q^2h$$
$$F_2 = 4p^2q^2(\tfrac{1}{2}h)$$
$$B_2 = 2 \times 2pq^3(-\tfrac{1}{2}d + \tfrac{1}{2}h)$$
$$P_2 = -q^4d$$

Summing these expressions and factorizing, we obtain

$$(p - q)(p + q)^3d + 2pqh(p + q)^2$$

As $p + q = 1$ (Units 9 and 10, equation 1), this simplifies to

$$(p - q)d + 2pqh$$

Now, the total variance of a BIPs generation is given by $\sum (x^2) - (\sum x)^2$. Again this formula is being used for a model situation in which we are using proportionate frequencies set by Mendelian principles of inheritance. Thus, where $(p - q)d + 2pqh$ is our $\sum x$,

$$\sum (x^2) = (p^2 + q^2)d^2 + 2pqh^2$$

and

$$(\sum x)^2 = [(p - q)d + 2pqh]^2$$

Thus

$$\sum (x^2) - (\sum x)^2 = (p^2 + q^2)d^2 + 2pqh^2 - [(p - q)d + 2pqh]^2$$
$$= 2pq[d + (q - p)h]^2 + 4p^2q^2h^2$$

* We have deliberately omitted the ms from the expressions for the means of the various families in order to simplify the subsequent algebra. It turns out that the ms all cancel.

For many-gene differences, the total heritable variance is the sum of the series of such items from each gene pair. Thus,

$$\sum 2pq[d + (q - p)h]^2 + \sum 4p^2q^2h^2$$

The random-mating form of D, written as D_R, is equal to $\sum 4pq[d + (q - p)h]^2$ in terms of a single-gene difference. For differences in many genes, D_R is equal to the sum of the series of such items from each gene. Thus,

$$D_R = \sum 4pq[d + (q - p)h]^2$$

This relationship was chosen because where $p = q = 0.5$, D_R reduces to $\sum(d^2)$, which is of course equal to D.

The random-mating form of H, written as H_R, is equal to $16p^2q^2h^2$ in terms of a single gene. For many genes, H_R is again the sum of the series of such items from each gene.

Thus,

$$H_R = \sum 16p^2q^2h^2$$

Again, where $p = q = 0.5$, H_R reduces to $\sum(h^2)$.

It is not too difficult to see that as $D_R = \sum 4pq[d + (q - p)h]^2$, and $H_R = \sum 16p^2q^2h^2$, the total heritable variance in a random-mating population is equal to $\frac{1}{2}D_R + \frac{1}{4}H_R$.

As we stated in the Study guide to Unit 12, you are not required to remember *any* of the algebra in Sections 12.3.1 and 12.3.2. The thing to remember is that D and H have been re-defined taking into account arbitrary allele frequencies and that, where $p = q = 0.5$, $D_R = D = \sum(d^2)$ and $H_R = H = \sum(h^2)$.

The important properties of D_R and H_R, are that D_R is the statistical additive or linear effects of the genes, and H_R is the statistical non-additive or non-linear effects of the genes. Unless, however, $p = q = 0.5$, then D_R contains some of the h effects, that is, dominance effects, of the genes, and H_R is correspondingly reduced.

12.4 D_R, H_R and the narrow-sense heritability

In Section 12.1, heritability was defined in terms of the parameters D, H and E for an F_2 generation.

QUESTION Define (a) the broad-sense and (b) the narrow-sense heritabilities in terms of D_R, H_R and E.

ANSWER (a) Broad-sense heritability,

$$h_B^2 = \frac{\text{total genetic variance}}{\text{phenotypic variance}}$$

$$= \frac{\frac{1}{2}D_R + \frac{1}{4}H_R}{\frac{1}{2}D_R + \frac{1}{4}H_R + E}$$

(b) Narrow-sense heritability,

$$h_N^2 = \frac{\text{additive genetic variance}}{\text{phenotypic variance}}$$

$$= \frac{\frac{1}{2}D_R}{\frac{1}{2}D_R + \frac{1}{4}H_R + E}$$

Let us consider an example. A random-mating population of the poppy, *Papaver dubium*, is raised in a randomized experimental design. The phenotypic variance of height is found to be 150 cm^2, the genetic variance is 90 cm^2 and the additive genetic variance 85 cm^2.

QUESTION Calculate the broad-sense and narrow-sense heritabilities.

ANSWER

$$h_B^2 = \frac{\text{genetic variance}}{\text{phenotypic variance}} = \frac{90}{150} = 0.60$$

$$h_N^2 = \frac{\text{additive genetic variance}}{\text{phenotypic variance}} = \frac{85}{150} = 0.57$$

So, by knowing D_R, H_R and E, both the broad-sense and narrow-sense heritabilities may be estimated. In Section 12.5 we shall examine how these components of the phenotypic variance may be calculated. At this stage, however, let us see how the size of the total genetic variance and the additive genetic variance depends on allele frequencies.

12.4.1 D_R, H_R and the total genetic variance in random-mating populations

To see how the total genetic variance is affected by changes in allele frequency we need to return to our expressions for D_R and H_R.

You should recall that

$$D_R = \sum 4pq[d + (q - p)h]^2$$

and

$$H_R = \sum 16p^2q^2h^2$$

The total genetic variance in a random-mating population is equal to

$$\tfrac{1}{2}D_R + \tfrac{1}{4}H_R = \sum 2pq[d + (q - p)h]^2 + \sum 4p^2q^2h^2$$

Using these expressions for the whole range of allele frequencies, we arrive at Figure 4, which reflects the contributions of the additive and dominance effects of the genes to the total heritable variance, when both d and h are equal to 1.

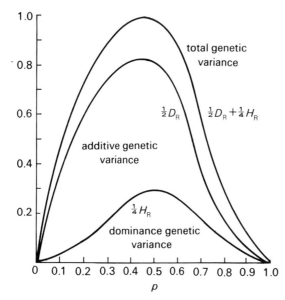

Figure 4 The contribution of allele frequency p to $\tfrac{1}{2}D_R$, $\tfrac{1}{4}H_R$ and the total genetic variance when $d = h = 1$.

The genetic variance is zero at the extremes where $p = 0$ or $p = 1$ because all the genotypes are uniform and, therefore, no genetic differences occur. As the allele frequencies move towards the middle of the range the genetic diversity and the genetic variance increase. D_R reaches its maximum at an allele frequency of just less than 0.5, and H_R reaches its maximum at 0.5.

At low allele frequencies nearly all the genetic variance is D_R in origin, H_R making little contribution. Conversely, at high allele frequencies the majority of the genetic variance is caused by H_R, and D_R contributes a relatively minor proportion.

You should recall from Section 12.1.2 that there is no correlation between the parental and offspring generations for genetic variance caused by dominance effects. Gene combinations in the parental generation are destroyed at meiosis and recombine at random in the progeny. Consequently, only phenotypic differences due to additive effects are expected to reappear unchanged in the progeny generation.

> QUESTION For the special case in which the A allele shows complete dominance, will the narrow-sense heritability, h_N^2, increase or decrease as a selection programme changes the frequency of the A allele from 0.1 to 0.9 in a population (see Fig. 4)?
>
> ANSWER The narrow-sense heritability is the proportion of the phenotypic variance that is additive genetic in origin. Consequently, from allele frequencies of 0.1 to 0.3, $\tfrac{1}{2}D_R$ and, therefore, h_N^2 increase and from allele frequencies of 0.3 to 0.9, $\tfrac{1}{2}D_R$ and h_N^2 decrease.

Remember that what is true for D and H is also true for D_R and H_R, which take arbitrary allele frequencies into account. And, of course, D_R and H_R are of paramount importance in predicting the possible response to selection in any breeding programme.

You should now attempt SAQ 4 (p. 580).

12.5 The narrow-sense heritability in random-mating populations

For any trait, the similarity between relatives can be estimated. Provided that there are no environmental similarities (and we shall discuss this later), the relatives' only source of similarity will lie in the possession of common genes.

This genetic covariance exists because the additive component of genes that are identical by descent causes the same effects in related individuals. In Section 12.1.2, we stated that only the additive effects of the genes, that is, the relative proportion of the increasing and decreasing alleles, will lead to similarities between the parental and offspring generations. So let us prove this by deriving the parent–offspring covariance!

12.5.1 The parent–offspring covariance

To make the statistics a little easier, we shall use the F_2 and BIPs generations to illustrate the relationship that exists between parents and their offspring. We do emphasize, however, that the same relationship holds true when we consider random-mating populations, where p and q take arbitrary values.

parent–offspring covariance

The covariance (Cov) is the sum of products of x and y (SP(x, y); see *STATS*, Section ST.10.2) divided by the degrees of freedom, $n - 1$:

$$\text{Cov}(x, y) = \frac{\text{SP}(x, y)}{n - 1} = \frac{\sum (x - \bar{x})(y - \bar{y})}{n - 1}$$

However, when we are dealing with a model situation using proportionate frequencies—set by Mendelian principles of inheritance—that sum to unity, then

$$\text{Cov}(x, y) = \sum (x - \bar{x})(y - \bar{y})$$

where x is the mean of the F_2 parents
  y is the mean of their biparental progeny
  $\bar{x}$ is the mean of the F_2 generation as a whole
  $\bar{y}$ is the mean of the BIPs generation as a whole.

Now, work through the following table. For example, take the mean of the F_2 parents for the P_1 family type. The mid-parent value for a cross between AA and AA is $\frac{1}{2}[(m + d) + (m + d)] = m + d$. Similarly, the mean of the biparental progeny is also $m + d$, only AA progeny being produced. Now, if we consider the mean of the F_2 parents for the F_1 family type, that is, AA and aa, this is given by $\frac{1}{2}[(m + d) + (m - d)] = m$. Similarly, the mean of the biparental progeny must be equal to $m + h$ because all the individuals are of Aa type. The remainder of the table is built up in a similar manner.

type of biparental family	P_1	B_1	F_1	F_2	B_2	P_2	generation mean
frequency	$\frac{1}{16}$	$\frac{1}{4}$	$\frac{1}{8}$	$\frac{1}{4}$	$\frac{1}{4}$	$\frac{1}{16}$	
mean of F_2 parents	$m + d$	$m + \frac{1}{2}d + \frac{1}{2}h$	m	$m + h$	$m - \frac{1}{2}d + \frac{1}{2}h$	$m - d$	$m + \frac{1}{2}h$
mean of BIPs	$m + d$	$m + \frac{1}{2}d + \frac{1}{2}h$	$m + h$	$m + \frac{1}{2}h$	$m - \frac{1}{2}d + \frac{1}{2}h$	$m - d$	$m + \frac{1}{2}h$

Consider the P_1 family type again,

P_1 parent

$$(x - \bar{x}) = (m + d) - (m + \frac{1}{2}h) = d - \frac{1}{2}h$$

P_1 BIPs

$$(y - \bar{y}) = (m + d) - (m + \frac{1}{2}h) = d - \frac{1}{2}h$$
$$(x - \bar{x})(y - \bar{y}) = (d - \frac{1}{2}h)(d - \frac{1}{2}h) = (d - \frac{1}{2}h)^2$$
$$= d^2 + \frac{1}{4}h^2 - dh$$

But only $\frac{1}{16}$ of the BIPs are of P_1 family type and they contribute $\frac{1}{16}d^2 + \frac{1}{64}h^2 - \frac{1}{16}dh$ to the covariance.

(You may like to try to work out the covariance for the remaining five family types, that is B_1, F_1, F_2, B_2 and P_2.)

To summarize: the covariance for the six family types is:

$$P_1 = \tfrac{1}{16}d^2 + \tfrac{1}{64}h^2 - \tfrac{1}{16}dh$$
$$B_1 = \tfrac{1}{16}d^2$$
$$F_1 = \qquad - \tfrac{1}{32}h^2$$
$$F_2 = \qquad 0$$
$$B_2 = \tfrac{1}{16}d^2$$
$$P_2 = \tfrac{1}{16}d^2 + \tfrac{1}{64}h^2 + \tfrac{1}{16}dh$$

Which sums to $\frac{1}{4}d^2$.

So, the covariance of the F_2 parents with their BIPs is equal to $\frac{1}{4}d^2$, which in terms of many genes is equal to $\frac{1}{4}D$, or for arbitrary allele frequencies is equal to $\frac{1}{4}D_R$.

Let us now see how this relationship may be used to calculate the narrow-sense heritability h_N^2.

12.5.2 The parent–offspring regression

In any random-mating population, the total phenotypic variance is equal to $\frac{1}{2}D + \frac{1}{4}H + E$ and the parent–offspring covariance is equal to $\frac{1}{4}D$. It should, therefore, be possible to find a relationship between these two expressions that gives us the narrow-sense heritability, h_N^2.

parent–offspring regression

> QUESTION What is the relationship between the total phenotypic variance, the parent–offspring covariance and the narrow-sense heritability?
>
> ANSWER The total phenotypic variance $= \frac{1}{2}D + \frac{1}{4}H + E$. The parent–offspring covariance $= \frac{1}{4}D$.

And the narrow-sense heritability $= \dfrac{\frac{1}{2}D}{\frac{1}{2}D + \frac{1}{4}H + E}$

$$\frac{\text{covariance of parents and offspring}}{\text{phenotypic variance of the parental generation}} = \frac{\frac{1}{4}D}{\frac{1}{2}D + \frac{1}{4}H + E} = \tfrac{1}{2}h_N^2$$

In practice, the simplest way to calculate the narrow-sense heritability is by regression analysis, where the y variable is the progeny family mean and the x variable is the value of one parent or the mean of both parents.

The regression coefficient,

$$b = \frac{\sum (x - \bar{x})(y - \bar{y})}{\sum (x - \bar{x})^2}$$

where

$$\frac{\sum (x - \bar{x})(y - \bar{y})}{n - 1} = \text{the covariance of parents and offspring}$$

and

$$\frac{\sum (x - \bar{x})^2}{n - 1} = \text{the variance of the parental generation}$$

Then,

$$b = \frac{\dfrac{\sum (x - \bar{x})(y - \bar{y})}{n - 1}}{\dfrac{\sum (x - \bar{x})^2}{n - 1}} = \frac{\sum (x - \bar{x})(y - \bar{y})}{\sum (x - \bar{x})^2}$$

$$= \frac{\text{sums of products of } x \text{ and } y}{\text{sums of squares of } x} = \frac{\text{SP}(x, y)}{\text{SS}(x)}$$

the $n - 1$ term being common to both numerator and denominator.

The regression of the offspring mean against one parent is equal to

$$\frac{\frac{1}{4}D}{\frac{1}{2}D + \frac{1}{4}H + E} = \frac{1}{2}h_N^2$$

and for arbitrary allele frequencies

$$\frac{\frac{1}{4}D_R}{\frac{1}{2}D_R + \frac{1}{4}H_R + E} = \frac{1}{2}h_N^2$$

The regression of the offspring mean against the mean of both parents is equal to

$$\frac{\frac{1}{4}D}{\frac{1}{4}D + \frac{1}{8}H + \frac{1}{2}E} = h_N^2$$

and for arbitrary allele frequencies

$$\frac{\frac{1}{4}D_R}{\frac{1}{4}D_R + \frac{1}{8}H_R + \frac{1}{2}E} = h_N^2$$

The statistical significance of any estimate of heritability must be assessed by calculating its standard error and applying a t-test. This is explained in STATS, *Section ST. 10.3, which we advise you to consult at this stage.*

QUESTION Consider some data for a large population of cattle that has not been inbred. Individuals were weighed to the nearest gram when they were 1 year old. When mature, 20 males were randomly mated and then their male progeny were weighed when 1 year old.

Calculate the narrow-sense heritability, h_N^2, for weight in this population. To make things easier we have given you the components of the estimate:

parent–offspring covariance = 450

variance of parents = 1 000

ANSWER The parent–offspring covariance is equal to

$$\frac{1}{4}D = 450$$

The variance of parents is equal to

$$\frac{1}{2}D + \frac{1}{4}H + E = 1\,000$$

$$\therefore \quad b = \frac{450}{1\,000} = 0.45$$

$$\text{and} \quad h_N^2 = 2 \times b = 0.90$$

Again, we really need to calculate the standard error of this estimate, but you will be able to do this in SAQ 5.

Now we come to the problem of correlated environmental effects. Provided both parental and progeny generations are not reared together in family units but are independently randomized over the environment, the covariance that exists between them can be due to genetic causes only.

In plants, this situation is very easy to achieve. If both the parental and progeny generations are grown independently and randomized, then no environmental correlations exist, and the covariance must be due to genetic effects only. In laboratory populations of animals, the same experimental design can ensure that there are no correlated environmental effects. In farm animals, this cannot always be assumed. Relatives may share a common environment, or they may not, depending on the management technique; for example, calves may be taken away from their mothers and bottle-reared independently or they may stay with their mothers. The important point is that correlated environmental effects can be overcome by appropriate management techniques.

Other relationships can be used in calculating the narrow-sense heritability, for example, sibs and half-sibs, but for the purpose of this Unit we shall restrict ourselves to the use of the direct method, that is, the parent–offspring regression.

12.5.3 Words of warning on heritability

The coefficient of narrow-sense heritability is not a constant but indicates only the relative proportion of the variance that is caused by additive gene effects in a particular population, at a specific moment and measured under a particular range of environmental conditions.

As changes in the environment are likely to lead to changes in gene expression, the narrow-sense heritability will change when measured in different environments. To take one extreme, if the environment were made completely uniform, then all the variation could be genetic in origin and the heritability could be high. At the other extreme, if the environment were completely heterogeneous, then much of the variation could be environmental in origin and the heritability could be very low.

Now, in many textbooks on genetics, authors say that narrow-sense heritability estimates are never comparative: they are a within-population statistic only. What do they mean by this and, equally, are they correct?

Narrow-sense heritability tells us the proportion of the phenotypic variance that will be of use to the plant and animal breeder. If a plant breeder wishes to assess material for a selection programme, he is likely to have obtained potential breeding material from many different populations. Provided all the experimental material is grown under the same environmental conditions, then any estimates of heritability are comparative. They tell the breeder how this material will perform in a selection programme within this set of environmental conditions. Clearly, plant and animal breeders try to assess their breeding material in the same or very similar environments to the ones in which the material will be grown or reared commercially.

Similarly, if a breeder is interested in assessing different characters within the same population or the same characters in different populations, then any estimates of heritability are directly comparable, but only if all the material has been grown under the same set of environmental conditions.

Conversely, if a plant breeder in Monmouth reports a population of perennial rye grass as having a heritability of 0.9 (dry weight), and at the same time a breeder in Glasgow working with the same species reports a heritability of 0.2, comparisons between them could be misleading because we would be comparing different populations measured in different environments.

Heritability is not a constant. It can change with the environmental conditions under which the material has been grown. However, if all the material, irrespective of population differences, has been assessed in the same range of conditions, such estimates are directly comparable.

In Unit 13 we shall discuss the application of biometrical techniques to ecological genetics. When dealing with ecological genetics great care must be taken when extrapolating from estimates of heritability. In plant and animal breeding programmes heritability estimates can be made in the same range of environments in which selection is to be applied, but in ecological genetics many of the characters of ecological interest are assessed in entirely different environments from those in which selection has been and is taking place. We shall however, discuss this point in greater detail in Unit 13.

You should now attempt SAQ 5 (p. 580).

12.6 Hybridization, mutation and alien gene pools as sources of new genetic variation

Plant and animal improvement depends ultimately on the availability of suitable genetic variation. To end this Unit we shall briefly examine three of the methods by which new genetic variation can be introduced into the gene pools* of cultivated crop-plant species.

At this stage you are strongly advised to consult your Broadcast Notes *for TV programmes 'The Genetic Manipulation of Wheat' and 'Plant and Animal Breeding'.* As you will see from the Objectives for this Unit, the contents of these programmes relate directly to this Section.

12.6.1 Hybridization

Interspecific hybridization, that is, the mating of two different species, has led to the development of many new and useful variants in crop-plant species. The technique is, however, of very limited use with animals as specific pre-mating rituals and differences in size often make the technique impracticable.

* The *gene pool* represents the total number of allelic forms of all possible genes within a breeding population.

In the TV programme 'The Genetic Manipulation of Wheat', we saw how the evolution of the hexaploid bread wheat depended on the interspecific hybridization of three distinct species—*Triticum monococcum*, *Aegilops speltoides* and *Aegilops squarrosa*.

Although this particular hybridization occurred without much intervention by man, there are examples where artificial interspecific hybridization has led to the development of new crop plants. For example, *Triticale* is the interspecific hybrid between wheat, *Triticum*, and rye, *Secale*. This new cereal species, which you saw in the TV programme, is still very much in its developmental stages but is showing great potential.

Another method employing hybridization is intraspecific hybridization in which matings are initiated between different strains or breeds within the same species.

For example, consider the production of the barley variety Proctor, which we mentioned in the Introduction. The two parents of this variety are Plumage Archer and Kenya.

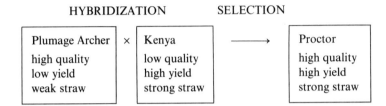

HYBRIDIZATION   SELECTION

Plumage Archer	×	Kenya	⟶	Proctor
high quality		low quality		high quality
low yield		high yield		high yield
weak straw		strong straw		strong straw

So, what are we doing when two lines are hybridized? We are effectively looking for groups that are genetically different from one another, and that on hybridization, will lead to segregation in the progeny and the release of new genetic variation. If we are lucky, the hybrids, or a selected strain from the hybrid generation, will possess the superior attributes of both parental types, as with the high quality, high yield and strong straw of Proctor barley.

Hybridization followed by selection is, therefore, one of the most practical and most common methods used to improve crop plants and animals.

12.6.2 Mutation breeding

In 1901, de Vries suggested that mutations could be used in breeding experiments and three years later even went so far as to suggest that X-rays could be used to increase the frequency of such mutations.

In 1928, Stadler in the United States demonstrated that mutations could be induced in varieties of barley. He was, however, pessimistic about the practical future of mutation breeding. The phenotypic mutants he had induced had been morphologically deformed, suffered from reduced viability and were sterile. Stadler's pessimism was undoubtedly transmitted to the plant breeders of this time and research into mutation breeding gradually came to a halt. From de Vries's original enthusiasm, mutation breeding had become a sadly neglected area of research.

After the Second World War and the advent of advanced radio-biological techniques, mutation breeding became a viable project once again, and is now at a sufficiently advanced stage to warrant evaluation.

All genetic variation must originally be derived from mutation. Granted that many mutations are disadvantageous, some at least must be conditionally advantageous and therefore retained in the gene pool of the species. An increase in the natural rate of mutation should lead to an increase in the production of new and useful variants.

Well, how is the rate of mutation increased and to what extent can it be increased? Many chemical mutagens are known, but the most potent agent in plants is ethyl methanesulphonate. This compound can produce up to 50 per cent of mutated offspring in successful treatments, and its advantage over ionizing radiations is that it does little harm to the chromosomes. Using ethyl methanesulphonate decreases the possibility of producing the gross physiological and morphological deformities that are very common with ionizing radiations.

Mutagens such as epoxides, sulphonic esters and ionizing radiations, can increase the natural rate of mutation by a factor of 500 or more. For example, the natural mutation rate of albino seeds in barley is in the order of 8×10^{-5} mutations per head per generation. Use of X- and γ-radiations produces a yield of 1.0×10^{-6} mutations per spike per generation per röntgen unit. This means that using a dose of 20 kR of X-rays we can obtain a mutation rate of 2×10^{-2} albino seeds per spike per generation.

Using such mutagenic agents, about 100 new varieties have been produced; these include 7 new varieties of wheat, 4 of rice, 11 of barley and 30 ornamental varieties.

We have no intention of cataloging new varieties and outlining their superiority, but it is useful to quote one or two examples.

In 1966 in Japan a commercial variety of Japonica-type rice was produced by γ-irradiation. The new variety, Reimei, is identical to its mother variety, Fujiminori, with the exception that it is resistant to lodging, or being blown over by the wind, and appears to be more resistant to changes in environmental conditions than its parental variety.

Induced mutations in the rice variety, Norin 8, produced individuals whose seeds yielded 100 per cent more than the parental type. There have been similar findings with induced mutations of barley in which both the protein content of the seed and the size of the seed have been increased by some 50 per cent above the parental type.

12.6.3 Gene pools and gene banks in plant breeding

All past and future advances in plant and animal breeding depend ultimately on the presence of genetic resources within the cultivated gene pool, and such resources can be divided into three well-defined classes.

The first area of new variation which we can examine is in the species closely related to our crop plants, and we can call these the wild gene pool. The wild gene pool consists of those wild forms that have close genotypic similarities with the agricultural species. Hybridization between the wild species and the agricultural cultivated varieties can lead to the introduction of new and useful variation to the cultivated gene pool.

The second class of gene pool has been termed the *alien gene pool*. Here, the variation is present in species belonging to the same biological genus as agricultural varieties, but is unavailable owing to the divergence of genomes or breeding barriers. Yet such species may well possess useful genes and in TV programme 'Genetic Manipulation of Wheat' we saw how the problem of introducing yellow rust-resistance from *Aegilops comosa* into the bread-wheat gene pool was overcome.

alien gene pool

Finally, there are the cultivated gene pools. Genetic variation occurs within and between cultivated varieties of crop plants and domestic animals, and intraspecific hybridization can lead to the expression and use of new variation. However, modern agricultural methods, that is, the use of a small number of uniform varieties of crop plants, has led to the erosion of this cultivated gene pool, less and less variation occurring as selection increases.

Up to 60 years ago, the old land varieties of many of our crop-plant species were genetically very variable. Paradoxically, with the advent of the Green Revolution, many of these varieties have been lost together with any alleles that may have been of use.

Recently, plant and animal breeders have become more and more aware of the threat of the rapid depletion and erosion of such genetic resources, and have initiated the establishment of gene banks, where germ-plasm can be stored in a viable state until its properties are needed in some hybridization.

We have only scraped the surface of the problem of genetic variation in plant and animal breeding in this Unit. If you are interested in this topic you are advised to read the two books recommended in the Bibliography and References Section.

12.7 Summary of Unit 12

In Unit 12 we have considered the parameters that are necessary for an understanding of the genetic principles of plant and animal breeding. The theoretical concepts have been expounded in the text and the practical aspects of plant and animal improvement are covered by the associated TV programmes.

The production of improved lines or varieties of plants and animals depends on two major concepts: first, that differences within a character occur in populations, that is, there must be variation in order to be able to exercise a selection, and, second, that the differences in any selected group of individuals must be at least in part genetically determined.

We also need to know in what way such characters are inherited. Is the variation all dominance? If it is, we cannot expect any similarity between the offspring and their selected group of parents. Only the additive effects of the genes are expected to cause resemblances between the parents and their offspring.

Appendix 1 List of symbols introduced in the text

V_g	the genetic variance
$V_g + V_e$	the phenotypic variance
h_B^2	the broad-sense heritability, $h_B^2 = V_g/(V_g + V_e)$
D	the additive genetic variance, $\sum (d^2)$
H	the dominance genetic variance, $\sum (h^2)$
σ^2	the population variance
E	the environmental component of the total phenotypic variance
h_N^2	the narrow-sense heritability
S_d	the selection differential
R	the response to selection
b	the regression coefficient
D_R	the statistical additive effects of the genes
H_R	the statistical non-additive effects of the genes
$\text{Cov}(x, y)$	the covariance of x and y
BIPs	the biparental progeny generation

Self-assessment questions

SAQ 1 (a) Derive the variance of an F_2 generation for a single-gene difference, in terms of the parameters d, h and E, and extend this derivation to include differences in many genes.

(b) Define the broad-sense heritability in terms of D, H and E in an F_2 generation.

(c) Calculate the broad-sense heritability in an F_2 generation for which the genetic variance of height in a variety of barley is $500 \, cm^2$, and the environmental variance is $250 \, cm^2$.

SAQ 2 (a) Derive the variance of the B_1 and B_2 generations for a single-gene difference and from this, demonstrate how $\frac{1}{2}D$ and $\frac{1}{4}H$ may be estimated.

(b) Define the narrow-sense heritability in terms of D, H and E.

(c) Given that the variance of height in a P_1 generation of barley is $100 \, cm^2$, calculate D and H, where $V(F_2) = 250 \, cm^2$, $V(B_1) = 180 \, cm^2$ and $V(B_2) = 200 \, cm^2$.

SAQ 3 A plant breeder is interested in producing a new line of barley that has a high yield of grain. The mean of his population is 280 g and he selects those individuals whose yield is over 300 g, the mean of the distribution being 320 g.

(a) Calculate the selection differential.

Now, assume that $V(P_1) = 120 \, g^2$, $V(F_2) = 500 \, g^2$, $V(B_1) = 320 \, g^2$ and $V(B_2) = 380 \, g^2$.

(b) Calculate the narrow-sense heritability, h_N^2.

(c) Calculate the response to selection in the progeny generation, and the mean of the progeny generation.

SAQ 4 (a) What are the relationships between

(i) d, D and D_R
(ii) h, H and H_R ?

(b) Under what circumstances does

$$D = D_R$$

and

$$H = H_R ?$$

(c) What is the total heritable variance of a random-mating population in which allele frequencies are arbitrary?

(d) Define the broad-sense and narrow-sense heritabilities in a random-mating population in terms of D_R, H_R and E.

SAQ 5 We give below data for body weight in a large flock of ducks that has not been inbred. Individuals were weighed at 40 days and 17 males were selected and mated randomly. Their male progeny were allowed to reach 40 days of age and at this stage they were also weighed.

(a) Calculate the narrow-sense heritability from the parent–offspring regression.

(b) Calculate the standard error of this estimate.

We have done the most time-consuming calculations for you.

$$\sum x = 14\,966, \text{ where } x \text{ refers to the male parents}$$
$$\sum y = 17\,353, \text{ where } y \text{ refers to the mean of the male offspring}$$
$$\sum (x^2) = 13\,375\,506$$
$$\sum (y^2) = 17\,752\,029$$
$$\sum (x, y) = 15\,319\,806$$
$$n = 17$$

Answers to ITQs

ITQ 1

$$h_{\text{B}}^2 = \frac{\text{genetic variance}}{\text{phenotypic variance}} = \frac{125}{125 + 25}$$

$$= \frac{125}{150} = 0.83$$

Note that the phenotypic variance is the sum of the genetic and the environmental variances. It is the observed or the total variance.

ITQ 2

$$h_{\text{B}}^2 = \frac{\text{genetic variance}}{\text{phenotypic variance}}$$

As the P_1 and P_2 generations do not undergo segregation, any variation must be environmental in origin. Taking the mean of the variance of these two generations as an estimate of E, that is, 15 cm^2, the genetic variance in the F_2 generation is equal to

$$35 \text{ cm}^2 - 15 \text{ cm}^2 = 20 \text{ cm}^2$$

$$h_{\text{B}}^2 = \frac{20}{35} = 0.57$$

That is, 57 per cent of the variation in the F_2 generation is genetic in origin.

ITQ 3

$$V(B_1) + V(B_2) = \tfrac{1}{2}D + \tfrac{1}{2}H + 2E = 850 \text{ cm}^2$$
$$V(P_1) = \phantom{\tfrac{1}{2}D + \tfrac{1}{2}H + 2} E = 300 \text{ cm}^2$$
$$\therefore \quad \tfrac{1}{2}D + \tfrac{1}{2}H = 250 \text{ cm}^2$$
$$V(F_2) = \tfrac{1}{2}D + \tfrac{1}{4}H + E = 500 \text{ cm}^2$$
$$\therefore \quad \tfrac{1}{2}D + \tfrac{1}{4}H = 200 \text{ cm}^2$$

As

$$\tfrac{1}{2}D + \tfrac{1}{2}H = 250 \text{ cm}^2$$

and

$$\tfrac{1}{2}D + \tfrac{1}{4}H = 200 \text{ cm}^2$$
$$\tfrac{1}{4}H = 50 \text{ cm}^2$$

and

$$H = 200 \text{ cm}^2$$

As

$$\tfrac{1}{2}D + \tfrac{1}{4}H = 200 \text{ cm}^2, \text{ and } \tfrac{1}{4}H = 50 \text{ cm}^2$$
$$\tfrac{1}{2}D = 150 \text{ cm}^2, \text{ and } D = 300 \text{ cm}^2$$
$$h_{\text{N}}^2 = \frac{\tfrac{1}{2}D}{\tfrac{1}{2}D + \tfrac{1}{4}H + E} = \frac{150}{500} = 0.30$$

That is, 30 per cent of the variation is additive genetic in origin.

ITQ 4

$$R = h_{\text{N}}^2 \times S_{\text{d}}$$
$$R = 0.83 \times 4 = 3.32$$

Therefore, the mean will increase by 3.32 g per generation of selection.

In order to get from 10 g to 19 g we would require three generations of selection.

Answers to SAQs

SAQ 1 (*Objectives 2, 5 and 7*)

(a) To check your derivation, return to Section 12.1.1.

(b)
$$h_B^2 = \frac{\text{genetic variance}}{\text{phenotypic variance}} = \frac{\frac{1}{2}D + \frac{1}{4}H}{\frac{1}{2}D + \frac{1}{4}H + E}$$

(c)
$$h_B^2 = \frac{500}{500 + 250} = 0.66$$

SAQ 2 (*Objectives 2, 4, 6 and 7*)

(a) To check this derivation, return to Section 12.1.2.

(b)
$$h_N^2 = \frac{\text{additive genetic variance}}{\text{phenotypic variance}}$$
$$= \frac{\frac{1}{2}D}{\frac{1}{2}D + \frac{1}{4}H + E}$$

(c)
$$V(F_2) = \tfrac{1}{2}D + \tfrac{1}{4}H + E = 250 \text{ cm}^2$$
$$V(P_1) = \qquad\qquad\quad E = 100 \text{ cm}^2$$
$$\therefore \quad \tfrac{1}{2}D + \tfrac{1}{4}H \qquad = 150 \text{ cm}^2$$
$$V(B_1) + V(B_2) = \tfrac{1}{2}D + \tfrac{1}{2}H + 2E = 380 \text{ cm}^2$$
$$\therefore \quad \tfrac{1}{2}D + \tfrac{1}{2}H \qquad = 180 \text{ cm}^2$$

Subtracting
$$\tfrac{1}{2}D + \tfrac{1}{4}H \qquad = 150 \text{ cm}^2$$
$$\tfrac{1}{4}H \qquad = 30 \text{ cm}^2$$

and
$$H \qquad = 120 \text{ cm}^2$$

As $\frac{1}{2}D + \frac{1}{4}H = 150 \text{ cm}^2$ and $\frac{1}{4}H = 30 \text{ cm}^2$, $\frac{1}{2}D = 120 \text{ cm}^2$, and $D = 240 \text{ cm}^2$.
$$H = 120 \text{ cm}^2$$
$$D = 240 \text{ cm}^2$$

SAQ 3 (*Objectives 7 and 8*)

(a) The selection differential is the difference between the mean of the selected group of parents and the mean of the population as a whole, before selection was applied. Thus,
$$S_d = 320 - 280 = 40 \text{ g}$$

(b)
$$h_N^2 = \frac{\frac{1}{2}D}{\frac{1}{2}D + \frac{1}{4}H + E}$$

Now,
$$V(F_2) = \tfrac{1}{2}D + \tfrac{1}{4}H + E = 500 \text{ g}^2$$

and
$$V(P_1) = \qquad\qquad\quad E = 120 \text{ g}^2$$
$$\therefore \quad \tfrac{1}{2}D + \tfrac{1}{4}H \qquad = 380 \text{ g}^2$$
$$V(B_1) + V(B_2) = \tfrac{1}{2}D + \tfrac{1}{2}H + 2E = 700 \text{ g}^2$$
$$\tfrac{1}{2}D + \tfrac{1}{2}H \qquad = 460 \text{ g}^2$$

Subtracting
$$\tfrac{1}{2}D + \tfrac{1}{4}H \qquad = 380 \text{ g}^2$$
$$\tfrac{1}{4}H \qquad = 80 \text{ g}^2$$

If
$$\tfrac{1}{2}D + \tfrac{1}{4}H = 380 \text{ g}^2 \text{ and } \tfrac{1}{4}H = 80 \text{ g}^2$$

then
$$\tfrac{1}{2}D = 300$$
$$h_N^2 = \frac{\frac{1}{2}D}{\frac{1}{2}D + \frac{1}{4}H + E} = \frac{300}{500}$$
$$h_N^2 = 0.60$$

(c) $R = h_N^2 \times S_d$

$R = 0.6 \times 40 = 24$ g

Therefore, the response to selection in the progeny generation is equal to 24 g.

As R is measured as a deviation from the mean of the whole parental population before selection was applied, the mean of the progeny generation is equal to 24 g + 280 g = 304 g.

SAQ 4 *(Objectives 9, 10 and 11)*

(a) (i) $\quad D = \sum (d^2)$

$\qquad H = \sum (h^2)$

(ii) $D_R = \sum 4pq[d + (q - p)h]^2$

$\qquad H_R = \sum 16p^2q^2h^2$

D_R is the statistical additive or linear effects of the genes and H_R is the statistical, non-additive, effects of the genes. The important point is that D and H have been redefined taking into account arbitrary allele frequencies.

(b) Where $p = q = 0.5$, as in an F_2 generation, then

$$D_R = D = \sum (d^2)$$

and

$$H_R = H = \sum (h^2)$$

Unless $p = q = 0.5$, D_R contains some h effects, and H_R is correspondingly reduced.

(c) The total genetic or heritable variance in a random-mating population with arbitrary allele frequencies is equal to

$$\tfrac{1}{2}D_R + \tfrac{1}{4}H_R$$

$$h_B^2 = \frac{\tfrac{1}{2}D_R + \tfrac{1}{4}H_R}{\tfrac{1}{2}D_R + \tfrac{1}{4}H_R + E}$$

$$h_N^2 = \frac{\tfrac{1}{2}D_R}{\tfrac{1}{2}D_R + \tfrac{1}{4}H_R + E}$$

SAQ 5 *(Objective 12)*

(a) Now, $b = \dfrac{\sum (x - \bar{x})(y - \bar{y})}{\sum (x - \bar{x})^2} = \dfrac{\text{sum of products of } x \text{ and } y}{\text{sum of squares of } x}$

Let us work out the denominator first. The easiest way to calculate the sum of squares of x is to use the expression

$$\sum (x^2) - \frac{(\sum x)^2}{n}$$

Thus,

$$SS(x) = 13\ 375\ 506 - \frac{(14\ 966)^2}{17}$$

$$= 200\ 144$$

Now, the sum of products of x and y is easiest to calculate by using the expression

$$\sum (x, y) - \frac{(\sum x)(\sum y)}{n}$$

$$15\ 319\ 806 - \frac{(14\ 966)(17\ 353)}{17} = 43\ 042$$

$$SP(x, y) = 43\ 042$$

The regression coefficient, $b, = \dfrac{43\ 042}{200\ 144} = 0.215$

and

$$h_N^2 = 2 \times b = 0.430$$

(b) What about the standard error of b and h_N^2?

The standard error of b is given by

$$SE = \frac{SD(b)}{\sqrt{\sum (x - \bar{x})^2}}$$

where $SD(b)$ is the standard deviation of b, and the denominator is the square root of the sum of squares of x.

Now the variance of b, that is, s_b^2 is given by

$$\frac{1}{n-2}\left\{\sum(y-\bar{y})^2 - \frac{[\sum(x-\bar{x})(y-\bar{y})]^2}{\sum(x-\bar{x})^2}\right\}$$

We know that

$$\sum(x-\bar{x})^2 = 200\ 144$$

and

$$[\sum(x-\bar{x})(y-\bar{y})]^2 = 43\ 042^2$$

$$\sum(y-\bar{y})^2 = \sum(y^2) - \frac{(\sum y)^2}{n} = 38\ 700$$

and

$$s_b^2 = \frac{38\ 700 - \dfrac{43\ 042^2}{200\ 144}}{15} = 1\ 963$$

The standard deviation of $b = \sqrt{1\ 963}$

$$\therefore \quad SE(b) = \sqrt{\frac{1\ 963}{200\ 144}} = 0.099$$

$$\therefore \quad \text{the standard error of } b = 0.099$$

As h_N^2 is equal to $2 \times b$, then the standard error of h_N^2 is twice the standard error of b, that is, 0.198. Our heritability estimate is equal to 0.43 ± 0.198, which can now be tested by a t-test.

The null hypothesis is that the slope of the line does not differ from zero. Thus

$$t_{[n-2]} = \frac{b-0}{S(b)} = \frac{0.43}{0.198} = 2.17$$

Reference to Table 3 of *STATS* (on p. 57) shows that a t-value of 2.17, with $n-2$ degrees of freedom, that is $n = 15$, is significant at a level of probability greater than 0.10 and, therefore, the heritability estimate is meaningful.

Bibliography and references

1 General reading

Bowman, J. C. (1974) *An Introduction to Animal Breeding*, Institude of Biology, Studies in Biology, **46**, Arnold.

Lawrence, W. J. C. (1967) *Plant Breeding*, Institute of Biology, Studies in Biology, **12**, Arnold.

Acknowledgement

Grateful acknowledgement is made to K. Mather and J. L. Jinks for permission to use Figure 4, which has been redrawn from an illustration in *Biometrical Genetics*, 2nd edn, Chapman & Hall (1971).

13 Ecological and Evolutionary Genetics

S299
GENETICS

Contents

List of scientific terms used in Unit 13

Developed in this Unit	Page No.	Developed in this Unit	Page No.
allopatric speciation	598	*K* selection	607
carrying capacity of the environment	604	*K* strategy	607
directional selection	589	logistic growth of populations	604
disruptive selection	590	morph–ratio cline	594
ecotype	591	phenotypic plasticity	593
environmental heterogeneity	590	polycross	600
exponential growth of populations	603	*r* selection	607
G × E interaction	594	*r* strategy	607
gene–ecological differentiation	590	stabilizing selection	588
heavy-metal tolerance	599	sympatric speciation	599

Objectives for Unit 13

After studying this Unit you should be able to:

1   Define, recognize the best definition of, and place in the correct context, the items in the list of scientific terms opposite.
(ITQ 3)

2   Discuss the basis and measurement of continuous variation.
(SAQ 1)

3   Demonstrate your understanding of the role played by the environment and genetic differences in determining continuous variation, by defining and calculating the narrow-sense heritability, phenotypic plasticity and G × E interaction.
(SAQ 1)

4   Discuss the effects of natural selection on continuous variation in terms of stabilizing, directional and disruptive selection, and recognize the three classes of selection from the appropriate information.
(SAQ 1)

5   Discuss the effects of environmental heterogeneity on continuous variation either in terms of ecotype differentiation or clinal patterns of variation.
(SAQ 2)

6   Recognize and use selected equations defining population growth.
(SAQ 3)

7   Discuss r and K strategy and selection with reference to examples used in the text of the Unit.
(SAQ 4)

Study guide for Unit 13

The main effect of selection is to alter the frequency of alleles in a population and, from Units 9 and 10, we have seen that the speed of evolution may be calculated for a discontinuous variate. In a continuously distributed variate, however, it is not possible to calculate allele frequencies because we cannot deal with individual loci. In continuous variation, we have to resort to the use of means and variances to describe evolutionary processes. It is, therefore, essential to understand the concepts of heritability (Unit 12), before reading this Unit.

In Section 13.3 of the text and the TV programme, the evolution of heavy-metal tolerance in plants is described in detail. SAQ 2 puts you in the position of an ecological geneticist and asks how you would attempt to demonstrate that salt-tolerant races of the red fescue grass *Festuca rubra* are, in fact, genetically differentiated ecotypes. Read Section 13.3 carefully, as many of the problems you will come across in SAQ 2 are discussed with reference to heavy-metal tolerance.

In the final part of Unit 13 we consider the mathematical description of population growth. *You will be expected to remember and use only the formulae that are numbered.*

13.0   Introduction to Unit 13

In Units 9 and 10, Section 9.2.1, when considering a discontinuous character, we saw how selection altered the frequency of phenotypes within a population. For example, the frequencies of the two morphs of the moth *Biston betularia* were seen to change in response to pollution and differential predation. This moth has only two character classes, speckled white and melanic black. But what happens when we consider a continuously varying character? How can we describe it?

You should recall from the statistics text* and from Units 11 and 12, that when a continuously distributed character is being considered, the character must be described in terms of the mean, μ, and the variance, σ^2, of its distribution. Different populations are likely to have different distributions and, therefore, different means and variances.

As in Units 11 and 12, we need to ask certain very specific questions about the distribution of a character. For example:

1   How much of the variation is genetic in origin?

2   What will be the resemblance between parents and their offspring?

In other words, we need to know the broad-sense and narrow-sense heritabilities of the particular character, in the particular population, in order to ascertain whether or not the variation is genetically determined and likely to be of evolutionary importance.

In this Unit, we shall discuss characters such as flowering time, tolerance of heavy metals and tolerance of salt—characters that are all continuously distributed and of evolutionary importance.

We shall look at the ways in which natural selection can alter the frequency of phenotypes within natural populations of plants and animals, and examine the effects of environmental heterogeneity on continuous variation—a subject that is called ecological genetics.

We shall also discuss the way in which the narrow-sense heritability can be estimated in natural populations of plants and animals, and analyse the necessity for this operation in ecological genetic studies.

In the final part of Unit 13, two relatively new topics in evolutionary population genetics, r and K selection and r and K strategy, are dealt with.

13.1   The effect of selection on continuously varying characters

If a character is varying in its expression from one individual to another, then it is reasonable to assume that some level of expression will be more consonant with the requirements of the environment than the others. The individuals that possess the optimum value of the character will have a selective advantage over the other individuals whose range of expression falls outside the optimum value.

For ease of handling, we can distinguish three types of selection—stabilizing selection, directional selection and disruptive selection.

13.1.1   Stabilizing selection

Stabilizing selection occurs when the mean phenotype of the population coincides with the optimum phenotype for that character in any specific set of environmental conditions.

stabilizing selection

Phenotypes departing from the mean will be at a selective disadvantage, the greater the departure the greater the disadvantage. The effect of selection will be the elimination of the variation of the character and the stabilization of the mean because the

---

* The Open University (1976) S299 STATS *Statistics for Genetics*, The Open University Press. This text is to be studied in parallel with the Units of the Course. We refer to it by its code *STATS*.

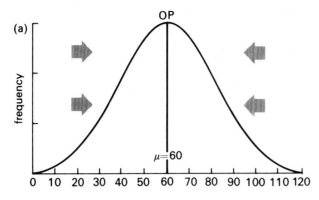

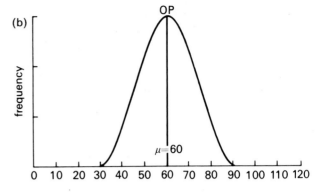

more variable offspring will be at a selective disadvantage and, therefore, the individuals that contribute most to the population will be those that give rise to the least variable progeny (see Fig. 1).

In 1952, Karn and Penrose made observations on the neonatal mortality of babies, that is, babies that die within 28 days of birth, and correlated these deaths with the birth weights. In all, 13 730 babies, 7 037 boys and 6 693 girls, born in London hospitals were observed. In Figure 2 you can see the distribution of neonatal mortality superimposed on that of birth weight. (For ease of presentation, the sexes have been combined.)

Figure 1 Stabilizing selection. This occurs where the optimum phenotype coincides with the mean phenotype. In Figures 1, 3 and 4 the direction of selection is indicated by the arrows; OP represents the optimum phenotype. (a) shows the distribution of the initial population; (b) shows the after-effects of the various types of selection on the distribution. The numbers on the abscissa scales are arbitrary. μ indicates the means of the phenotypes.

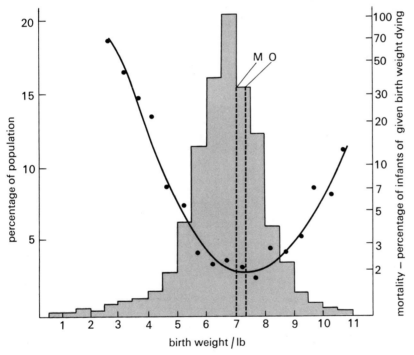

The rate of mortality is minimal at a weight approximating to the mean, which is about 7.5 lb (3.4 kg), and rises sharply on either side of it. The causes of death are no doubt different for over-large and under-sized babies, size being the most important factor in both antenatal care and delivery; but this is irrelevant to our present discussion. The effect of selection in this instance is to stabilize the birth weight.

Figure 2 Distribution of human birth weight (histogram) and the rates of neonatal mortality in various birth-weight classes (graph). M = mean birth-weight; O = lowest mortality.

13.1.2 Directional selection

One might expect that selection in the wild would in the main have a stabilizing effect—provided, of course, one assumed that most populations of organisms are relatively well adapted to their own specific environment.

But what happens if the optimum phenotype does not correspond to the mean of the population? Selection will favour some phenotype within the total range of the character, but this phenotype will be one that does not characterize the mean. In other words, selection will be directed against the mean, and under this *directional selection*, the population will be skewed towards either extreme of the range of the distribution (see Fig. 3).

directional selection

Most known instances of directional selection occur in laboratory experiments or plant and animal breeding programmes. The optimum phenotype is likely to lie outside the mean of the character's expression, and selection is directed towards a bigger or better or smaller individual than the one that characterizes the mean.

There are, however, interesting examples of directional selection in the wild and we shall examine one of these, DDT resistance in the mosquito (family Anophelinae).

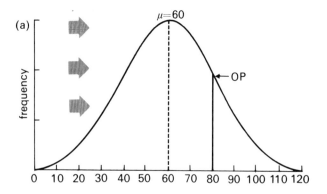 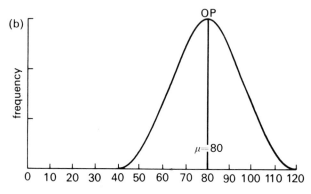

In the early 1960s in Bangkok, spraying DDT at a concentration of 0.005 parts per million (parts/10⁶) killed off 50 per cent of the mosquitos. In 1967, over 5.0 parts/10⁶ were required to achieve the same rate of control, and by 1969, 15 parts/10⁶ were required. Clearly, by spraying with DDT, man has inadvertently selected DDT-resistant strains of mosquito, until the population has become totally resistant.

Figure 3 Directional selection. This occurs where the optimum phenotype is not the mean.

13.1.3 Disruptive selection

The third class of selection may arise when a change in environment results in two or more extreme phenotypes being favoured at the expense of the mean. Under such *disruptive selection*, there are two or more optimum phenotypes, and this could lead to the establishment of a balanced polymorphism (see Fig. 4).

disruptive selection

With a discontinuous variate, it is easy to appreciate that by maintaining both phenotypes, corresponding to the genotypes *AA* and *aa*, within the population, the population will become dimorphic or polymorphic. In Units 9 and 10, we saw that such a balanced polymorphism could be maintained by *environmental heterogeneity*, each morph being adapted to mutually exclusive environmental conditions.

environmental heterogeneity

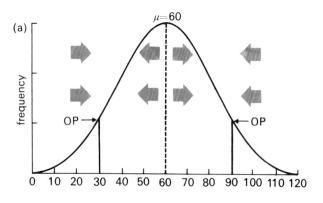 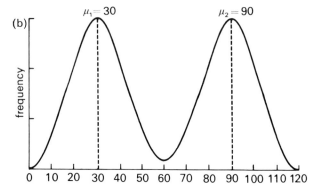

When continuous variation occurs under a range of environmental conditions, there is opportunity for the development of locally adapted and genetically differentiated sub-populations or races by way of disruptive selection. In Section 13.2 we shall explore the evolution of locally adapted races, a process that we can call *gene–ecological differentiation*.

Figure 4 Disruptive selection. This occurs where there are two or more optimum phenotypes.

gene–ecological differentiation

ITQ 1 In 1898, Bumpus measured nine different characters in the house sparrow, *Passer domesticus*, both in birds that had died as a result of a severe storm in New York and in birds that had survived.

For eight of these characters, Bumpus found that the surviving birds had measurements that tended to cluster around intermediate values, but the dead birds exhibited a great deal of variation within these eight characters. Some of the characters he looked at were body weight, the length of tail and the size of wings and beak.

Explain these results in terms of natural selection and, in particular, in terms of stabilizing selection.

The answers to the ITQs are on p. 611.

You should now attempt SAQ 1 (p. 609).

13.2   Plant evolution in a heterogeneous environment

Many habitats that are occupied by plants are highly heterogeneous. This means that within a single population, or within the species as a whole, local selection pressures may act as centres of disruptive selection.

Let us first consider disruptive selection on a major scale, that is, within the total range of a species.

13.2.1   Ecotypic differentiation

If we consider any species that has a wide ecological range, then certain groups of phenotypes within the total range of the species will be better suited or adapted than others to specific habitats. For example, in the grass, *Agrostis tenuis*, there are particular groups of phenotypes that occur in different soil conditions, and such groups are called *ecotypes:* calcicole ecotypes occur in high-calcium soils, calcifuge ecotypes occur in low-calcium soils and heavy-metal-tolerant ecotypes occur in soils contaminated by heavy metals.

ecotype

It has been demonstrated that these three ecotypes are both physiologically and genetically adapted to life in the three different sets of soil conditions, and the occurrence of ecotypes must be a consequence of different selection pressures in different habitats. In this case, selection is determined by the soil type.

QUESTION   Another example of adaptation to one type of soil occurs in the grass, *Andropogon*. Ecotypes of this grass have been found growing in soil containing uranium at concentrations that are normally toxic to most plant species. When seeds from normal populations and from populations growing on uraniferous soils were grown in contaminated soil, the results showed quite clearly that the seeds collected from the contaminated soils germinated and survived far better than seeds collected from normal populations. Do these results prove, however, that the ability to survive on toxic soils is genetically determined?

ANSWER   The fact that this experiment was performed on seed material tells us that the variation in uranium-tolerance has been inherited by the progeny and is, therefore, genetically determined.

Let us now consider a rather more extensive example of ecotypic differentiation in the rock rose, *Potentilla glandulosa*. In 1948, Clausen, Keck and Hiesey published data on the distribution of four ecotypes of *P. glandulosa* in the Sierra Nevada of California. The four ecotypes and some of their characteristics are shown in Table 1.

Table 1   Ecotypes of *P. glandulosa*

Ecotype	Approximate distribution (height)	Winter growth
typica	coastal ranges and foothills	grows all the year round
reflexa	274–1 830 metres	grows all the year round
hanseni	1 220–2 440 metres	dormant in winter
nevadensis	2 440–3 350 metres	dormant in winter

591

An overlap occurs in the ranges of altitude of *hanseni* and *reflexa*, but *hanseni* is restricted to wet alpine slopes and *reflexa* occurs on dry, steep, slopes. We can define an ecotype as a genetically differentiated sub-population or race that is restricted to a specific habitat—so it certainly appears that *hanseni* and *reflexa* are ecotypes, although we have as yet no evidence about whether or not they are genetically different. How could we establish whether or not the four ecotypes of *P. glandulosa* are restricted to specific habitats and whether they are genetically different?

In fact, Clausen, Keck and Hiesey took representative samples of the four ecotypes and propagated them vegetatively, that is, by clonal reproduction (see Units 9 and 10, Section 9.4.4). Clonal material from each plant was then transplanted into three experimental stations at different heights in the Sierra Nevada and a comparative performance chart was constructed, similar to Table 2.

Table 2 Comparative performance of the four ecotypes of *P. glandulosa*

Station and altitude	Stanford 33 metres	Mather 1 400 metres	Timberline 3 050 metres
typica	optimum growth	very small	dies
reflexa	very small	optimum growth	dies
hanseni	dormant in winter	optimum growth	rarely sets seed
nevadensis	barely survives	optimum growth	sets seed

From these results, we can see that the different ecotypes perform at their optimum when they are at an altitude most like that from which they came, that is, they are fittest in their own environment. Note that we are talking about survival fitness here rather than reproductive fitness. Thus, *typica* performs better at Stanford than the other two stations, and *nevadensis* performs better at Mather and at Timberline. The ecotypes that are planted in the environment most removed from their own respond very badly. For example, *hanseni* remains dormant in winter, even at Stanford, and *typica* and *reflexa* fail to grow at all at Timberline.

Thus, the four ecotypes of *P. glandulosa* appear to be suited to different altitudes and associated climatic conditions. We can call these ecotypes 'climatic ecotypes', because it is the climate that is selecting the ecotypic response of the species.

But has this proved that differences between the four ecotypes are genetically determined? In other words, are we sure that the variation within the characters that we are looking at is heritable and likely to be of evolutionary significance?

In the past, many ecological geneticists have assumed that if differences between ecotypes are retained when plants are taken from their normal habitats and grown under standard environmental conditions (i.e. taken into cultivation), such differences are genetically determined and are of evolutionary significance. The situation is not quite as simple as this, however. The retention of differences in cultivation no doubt suggests there is a genetic component to the variation, but it does not prove it, nor can it tell us how much is genetic. The only way to prove that such variations are heritable is to look at the parent–offspring regression and determine the narrow-sense heritability and, of course, the statistical significance of this estimate (see Unit 12, Section 12.1).

Clausen, Keck and Hiesey did look at the relationship between the parents and their offspring, and found that the characters that appeared to reflect levels of fitness were in fact heritable and likely to be of evolutionary significance.

You might, of course, ask why did people assume that differences retained in cultivation are genetically determined. The assumption that differences between different populations or different ecotypes would disappear if they were subjected to a common environment is a hangover from the early descriptive phase of ecological genetics. Only if such differences were genetically determined would they be retained in cultivation.

Let us analyse how and why such a thesis developed and, second, see what is fundamentally wrong with it.

In describing the differences between ecotypes we can only compare their phenotypic attributes. Only recently have ecological geneticists capitalized on the biometrical approach and started to determine the extent to which such variations are heritable.

The somewhat slow adoption of biometrical genetics is not as surprising as it may at first appear. A true understanding of heritability depends on the application of the principles that you learnt in Units 11 and 12, and many workers in the field failed to realize or understand the significance of heritability, preferring to remain at a descriptive level.

There is another reason. Suppose that two individuals of perennial rye grass, *Lolium perenne*, are taken as representatives of two distinct populations. Plant G_1 is taken from a roadside verge where all the plants are tall, and plant G_2 is taken from a grazed pasture where all the plants are dwarf. Both plants are grown in a common environment, and after 2 years their heights are compared with their original height.

	G_1	G_2
initial environment	80 cm	20 cm
common environment	100 cm	40 cm

Both plants have responded phenotypically by increasing their height in the common environment. Plant G_1 has increased its phenotypic expression from 80 to 100 cm, and plant G_2 from 20 to 40 cm.

This ability of plants to change their phenotypic expression is termed the *phenotypic plasticity*. We can calculate the plasticity of height, in this instance, or of any other character, by subtracting the phenotypic value in the original environment from the phenotypic value in the common environment and dividing by two. We divide by two because we are comparing the two phenotypes of a single genotype. Thus,

phenotypic plasticity

$$P_{G_1} = \frac{100 - 80}{2} = 10$$

$$P_{G_2} = \frac{40 - 20}{2} = 10$$

where P_{G_1} is the plasticity of plant G_1 and P_{G_2} is the plasticity of plant G_2.

Clearly, both plants have responded equally to the common environmental conditions, that is, they are of equal plasticity.

QUESTION Can you suggest why we would expect to find short individuals in a grazed pasture?

ANSWER In a grazed pasture, any tall individuals would be likely to be eaten more often. Now, in this situation, two types of short individual may occur. First, genetically determined short individuals, and second, phenotypically determined short individuals, that is, those that have been grazed down by herbivores.

In this example, it appears as though pasture plant G_2 is genetically short. If it were phenotypically short, we would expect it to grow taller in the absence of grazing. But does this observation prove that the variation in height between the two individuals is genetically determined? Before trying to answer this question, look at Table 3.

Table 3 **The plastic response of two characters in linseed,** ***Linum usitatissimum*, to changes in density conditions, that is, spacing between plants**

Character	High density	Low density
capsules (seed pods)	6	77
number of seeds per capsule	8	9

QUESTION What observations can you make about the two characters from the data given in Table 3?

ANSWER The two characters are showing different responses to changes in density, that is, they are showing different degrees of plasticity. Obviously, the number of capsules per plant is highly plastic, whereas the number of seeds per capsule is relatively stable.

If different characters within the same individual show differences in plasticity, it is only reasonable to assume that the same character in different individuals may show differences in plasticity over the same range of environmental conditions.

When differences in plasticity occur between different individuals for the same character over the same range of environmental conditions, we call this *genotype × environmental interaction*, which is denoted by G × E.

G × E interaction

A simple way of looking at G × E interaction, is to consider a two-way analysis of variance (an extension of one-way analysis of variance discussed in *STATS*, Section ST.9.1). For example, take two genotypes G_1 and G_2 growing in two different environments.

environment	genotypes	
	G_1	G_2
x	A_1	B_1
y	A_2	B_2

The analysis of variance has three degrees of freedom, genetic effects, environmental effects and genetic × environmental effects. Now, if different individuals are taken from different populations and grown in common environmental conditions, it is naïve to assume that both individuals will respond equally, that is, that there will be no G × E interaction. Of course, we can determine whether or not G × E is occurring by carrying out a suitable experiment.

If differences are retained in cultivation, then nine times out of ten we may be sure that there is some genetic component determining the variation in the character, but how much we cannot say.

To reiterate our original statement: the only way to prove that differences are genetically determined and, therefore, likely to be of evolutionary significance, is to determine the narrow-sense heritability of such characters, and to assess the statistical significance of any estimate.

13.2.2 Clinal variation

The term cline (Units 9 and 10, Section 9.3) was proposed by Huxley in 1938 to signify a character gradient. When ecological changes are gradual, this can often be reflected in a gradual change in the appearance of an organism over its entire range, rather than an abrupt change at a certain point where one type gives way to another.

cline

An organism may belong to any number of clines because a cline refers to a gradient for a single character only. For example, a cline may be determined by ecological changes, that is, it is an ecocline, and superimposed on this cline we may have one determined by geographical changes, that is, a topocline. Of course, any one, or any combination, of the multitude of the components within these two very general spheres could result in independent clinal variation patterns.

You might imagine that a cline is such a mass of unrelated soil types, climatic or geographical variables, that it is impossible to unravel the determining components of the environment. Fortunately, however, the determining variables are usually quite discernible.

Even a discontinuous variate may show a clinal pattern of variation. Indeed, in one of its earliest uses, the term cline was applied to a series of populations in which the frequencies of two distinct morphological forms of the guillemot changed from south to north up the coast of Great Britain.

When changes in the frequency of morphs follow a clinal pattern, this is often referred to as a *morph–ratio cline*. Let us consider an example. We return to industrial melanism in the peppered moth, *Biston betularia*, which was discussed in some detail in Units 9 and 10, Section 9.7.1. In the 1960s, Bishop, in Liverpool, released adult populations of *B. betularia*, consisting of 50 per cent melanic (*carbonaria* + *insularia*) and 50 per cent speckled-white (*typica*) morphs, in two very different localities. In North Wales, 60 per cent of the melanic morphs were taken by predators, and in Sefton Park, in the middle of Liverpool, only 45 per cent of the melanic morphs were taken.

morph–ratio cline

QUESTION How can you interpret these findings?

ANSWER The observations show quite clearly that the melanic morph was at a 15 per cent visual disadvantage in North Wales, which is relatively unpolluted compared with the middle of an industrial city. This result further emphasizes the selective advantage of cryptic colouration in industrial areas.

Using a similar approach, Bishop took samples from populations of *B. betularia* in Liverpool and a series of sites leading out into North Wales. The results of these observations are shown in Table 4.

Table 4 The relative frequencies of the two morphs of *B. betularia*

Site	Melanic morph (%)	Speckled-white morph (%)
Liverpool	97	3
Wirral*	93	7
Caldy	87	13
Colwyn Bay	11	89
Bangor	3	97

* This entry represents the mean of many samples taken throughout the Wirral peninsula.

ITQ 2 A frequency-distribution chart can be constructed by representing the frequencies of the two morphs at each of the sampling sites in standard rectangules and shading in the appropriate area with different colours.

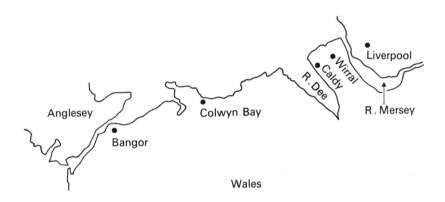

Figure 5 Map showing the sites where Bishop took samples of *B. Betularia*.

Using the method outlined, construct a frequency-distribution chart of the two morphs above the map in Figure 5, and comment on this distribution.

The two forms of moth are distinguished by alleles of a single gene, the melanic morph being dominant to the speckled-white form. Therefore, the gradient, or cline, from Liverpool to North Wales can be considered in terms of a change in allele frequencies related to distance. But clines are often observed in characters showing truly continuous patterns of variation, and here again we can think in terms of a gradient in allele frequencies. In many cases, however, the techniques required to study the genetic situation are too complicated and, in these instances, the cline represents a change in phenotypic frequencies only.

To illustrate clinal patterns of variation in plants we shall take some further data of Clausen, Keck and Hiesey's on the distribution of *Achillea lanulosa*, which shows a clinal variation pattern in height across the Sierra Nevada. They took plants from various locations, grew them in a common environment at Stanford and measured their height, producing a graph similar to the one shown in Figure 6.

QUESTION Why did Clausen *et al.* need to grow all the plants in a common environment?

ANSWER The variation they observed in the height of the plants could have been a plastic response to altitude rather than a specific gene–ecological differentiation. As it was not possible to perform a genetic analysis on this material, it was essential to compare the different individuals in a common environment. Of course, this assumes that there would be no G × E inter-action, which as we have seen may be an oversimplification. However, if such differences are retained in cultivation, this suggests a genetic component in their determination.

In Figure 6, the height of the plants from different habitats has been plotted against altitude, and we can see a direct and continuous relationship between the two. As the altitude increases, the height of the plants decreases, the height following the variation in the altitude. It is possible that the relationship between the height of the

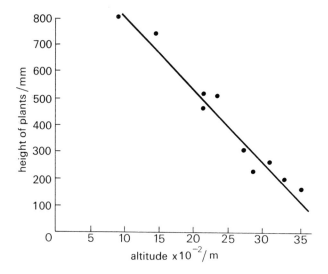

Figure 6 Height of plants of *Achillea lanulosa* grown in a uniform environment, plotted against the original altitude at which they normally grow. The graph shows a straight-line relationship between the height of the plants when grown in uniform conditions and the altitude at which the same plants normally grow (i.e. from which they were originally taken).

plant and the altitude is determined by a balance between the effects of competition and climatic conditions, especially exposure. A plant with a prostrate habit may be at a competitive disadvantage at low altitudes, whereas at higher altitudes the taller plants may suffer severely from exposure. If this is correct, then at any point in the cline there is an optimum height, which is determined by the balance between competition and the degree of exposure. This relationship is pure conjecture at the moment and further experimentation would be necessary to prove it.

We have seen that over a wide range of distribution, gene–ecological differentiation may result in either the evolution of specifically adapted discontinuous ecotypes in response to discontinuous environmental variables, or clinal patterns of variation, in which differences overlap, leading to gradual and continuous character gradients in response to gradual and continuous changes in the environment.

But one might well ask whether such processes could occur over relatively short distances? Let us first look at ecotypic differentiation. If we think of a sharp boundary between two exclusively different habitats, then disruptive selection may have resulted in the divergence of the original population into two genetically different ecotypes, and this may well be at distances of less than 1 or 2 metres. But what happens in clinal patterns of variation? Can clines occur over correspondingly short distances? The answer to this question is yes. The experimental evidence from several plant species suggests that the distance separating the two extreme ends of a cline may be smaller than 50 metres. Snaydon has recently examined clinal variation patterns in the grass, *Anthoxanthum odoratum*, at the Park Grass Experiment at Rothamsted Experimental Station.

As long ago as 1886 experimental plots were established by Lawes and Gilbert to investigate the long-term effects of constant application of fertilizer and lime to an otherwise uniform area of grassland. As one might expect, the botanical composition of the plots treated with different fertilizers has altered greatly since that time. Of equal interest has been the change in the general growth of the plants.

Now the two plots in which Snaydon was interested were identical except for the amount of lime they received: one was limed in 1903 and subsequently every 4 years, and the other received no lime at all. The boundary between the two plots was less than 50 cm, and yet distinct morphological and physiological differences have been revealed that correlated with the different environments of the two liming regimes. More recently, Davies has produced evidence to suggest that such differences are indeed genetically determined.

So, we can have population differentiation in distances of less than half a metre. Can we go any further? Following some earlier work of Bradshaw with various species of grass, Cook worked on the Californian poppy, *Eschsholotzia californica*, and analysed the genetic variation shown by many characters, such as stamen length, seed weight and seed length, in over 30 populations. He found that the pattern of variation in an individual character was related to the micro-geographical variations in the local environment. Each character was independent of the others and was involved in a separate cline and, in this way, the population was involved in a series of clines of varying slopes, interspersed with sharp discontinuities corresponding to major environmental discontinuities. Each individual was, therefore, at the optimum for every environmental variation and was specifically and genetically adapted to its own particular micro-site. Bradshaw and Cook have termed this situation a 'graded patchwork effect'.

We have, therefore, a situation in which localized patterns of micro-geographical variations result in a correspondingly localized pattern of differentiated individuals that are in part genetically determined, each individual being adapted to its own particular micro-environment.

13.2.3 The consequences of disruptive selection

In the previous Section, we saw that environmental heterogeneity and the associated disruptive selection resulted in either the evolution of distinct ecotypes or clinal patterns of variation. Individuals in ecotypes or clines are likely to be genetically differentiated and this may lead to the formation of races, which is the initial and crucial step in speciation.

As a result of accumulating genetic differences, other differences, both morphological and physiological, will follow until the races are clearly distinguishable from one another and are formally classified as sub-species.

Sub-species are interfertile (that is, they can breed freely) and their ultimate fate depends on whether or not reproductive isolation subsequently develops. If the environmental heterogeneity continues and becomes more pronounced, resulting in a greater divergence, and the sub-species become so genetically different that they are no longer interfertile, then we see the formation of a new species. If, on the other hand, the sub-species continue to share a common gene pool, then they may either merge again, or persist more or less indefinitely, depending on the balance between migration and selection.

QUESTION In Units 9 and 10, we saw that the genetic structure of populations could be altered by four major processes. What are these?

ANSWER The major processes that may alter the genetic structure of populations are mutation, selection, genetic drift and migration.

QUESTION What is meant by migration in this context, first, in animals and, second, in plants?

ANSWER In animals, migration means that members of genetically different populations join other populations and interbreed. You should recall from Units 9 and 10, Section 10.4.1 that the change in allele frequency due to migration is a function of the number of generations during which migration has been taking place, the rate of migration and the allele frequencies of the migrant and recipient groups.

In plants, migration has been termed 'gene flow', or more correctly pollen flow, because it is the pollen grains that carry the male gametes.

Evidence suggests that the pollen flow occurs in a *leptocurtic distribution* (Fig. 7, p. 598), showing a high rate over short distances, with a rapid fall over 8 metres, resulting in a low residual rate over large distances. This distribution is the same for both wind-borne pollen and pollen carried by insects.

leptocurtic distribution

Let us see how this information was obtained. Using the recessive white-based genotypes of perennial rye grass, *Lolium perenne*, Griffiths established plots of white-based plants increasing in distance from a plot of red-based genotypes. He then allowed nature to take its course and determined the percentage of red-based plants in seed collected from the white-based plants. A graph similar to the one shown in Figure 7 was constructed.

As you can see, the percentage of contamination from the red-based plants, which must be due to pollen flow, is at its highest over short distances, decreases rapidly over 8 metres and results in a residual rate of contamination over longer distances of some 2 per cent.

The problem we now face is that if two closely adjacent populations of plants are the result of disruptive selection pressures, both being genetically adapted to their own specific environmental conditions, then surely migration, or gene flow, between the

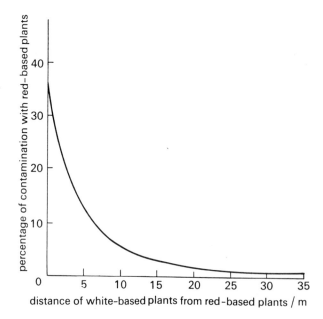

Figure 7 The effect of isolation distance on the contamination of plots of white-based plants of *Lolium perenne* by red-based plants.

two populations is likely to counterbalance the divergent selection pressures. How can populations diverge in the face of the gene flow between them?

This situation is perhaps one of the most controversial issues concerning speciation. If the balance between migration and divergent selection pressures is in favour of migration (that is, selection pressures are weak), then the two sub-populations will share a common gene pool and are likely to converge and become similar. If, on the other hand, selection pressures in the two sub-populations are greater than the migration rate, then there will be considerable differentiation of the two sub-populations.

Speciation requires some method of reproductive isolation. Controversy arose about whether it is necessary to have a physical barrier to migration before reproductive isolation can be initiated. Putting this the other way round, is a physical barrier a prerequisite for speciation, so that by effectively eliminating migration, the sub-populations could diverge genetically as a result of mutation, differential selective forces and genetic drift?

There are indeed instances of speciation as a result of geographical barriers to migration, and we refer to these as *allopatric speciation*. The most convincing **allopatric speciation** demonstration of the role of geographical barriers in speciation is found in island groups where barriers are clearly defined. For example, in the Hawaian and Galapagos archipelagos, each island can accumulate many different species from a single ancestral type.

In these cases we can identify three specific phases that lead eventually to allopatric speciation:

1   The initial invasion by a random sample of the ancestral population.

2   The isolation of the island population.

3   The differentiation of populations on separate islands.

For example, a single type of finch that colonized the Galapagos islands some time in the past has evolved into 13 distinct species.

There appears, however, to be some inconsistency in our logic, for birds can fly to islands and establish new populations, and yet we suggest that island populations are geographically isolated. This can be resolved by comparing the relative rate of immigration needed to colonize an island with that required to provide effective levels of gene flow. The infrequent dispersal of a few individuals to an island is likely to be sufficient to establish a new population. Once this population has grown, however, the occasional arrival of a few immigrants from a different stock would have little effect on the gene pool.

In contrast to allopatric speciation, *sympatric speciation* may arise as a result of ecological preferences. What do we mean by this? We have seen that disruptive selection and environmental heterogeneity result in the evolution of distinct ecotypes or clinal patterns of variation, in which the individuals are genetically suited to their own micro-environment. You should recall that if the balance between migration among various populations and the divergent selection pressures is tilted in favour of selection, then local differentiation could occur in the sub-populations, which may eventually lead to the accumulation of genetic differences to such a point that reproduction between the divergent groups is no longer possible. At this stage, sympatric speciation has occurred.

sympatric speciation

> **ITQ 3** (a) Define 'ecotype'. Illustrate your definition by reference to examples used in this Unit.
>
> (b) Define allopatric and sympatric speciation.
>
> (c) The gene flow between two closely adjacent populations of the grass, *Festuca ovina*, is known to be at a high level. What will be the long-term outcome of the situation in which:
>
> (i) the two populations are occupying slightly different environments, but the rate of gene flow is greater than any possible divergent selection pressures within the populations?
>
> (ii) the two populations are occupying mutually exclusive environments?

13.3 The evolution of heavy-metal tolerance in plants

The evolution of *heavy-metal tolerance* in plants is now one of the most widely investigated and quoted examples of evolution; it brings together many of the concepts we have previously discussed. This subject is also treated in detail in the TV programme on heavy-metal tolerance.

heavy-metal tolerance

13.3.1 The environment

The mine in the photograph in Figure 8 is typical of many derelict, heavy-metal mining areas, where the waste, or spoil, as it is known, has been dumped in haphazard heaps. Such waste often contains as much as 10 per cent of lead, zinc, copper or arsenic, which at these concentrations are extremely toxic to both plants and animals, causing protein precipitation, toxaemia and death.

Figures 8, 9 and 10 are opposite p. 600.

If only for aesthetic reasons it is fortunate that a limited number of plant species do, in fact, colonize and reproduce in such habitats; it is the occurrence of these plants that we shall be discussing.

The species with which we shall be concerned are the grasses, *Agrostis tenuis*, *Anthoxanthum odoratum* and *Festuca ovina*, because these are common members of the mine-site flora and are the ones on which most of the research has been performed.

13.3.2 The story

Suppose we have a population of the grass *A. tenuis* spread over an area of some 100 m^2, in which there is a small area of waste contaminated by heavy metals. We would expect disruptive selection caused by environmental differences to lead to the evolution of two distinct ecotypes: first, non-tolerant ecotypes that colonize un-contaminated land adjacent to the waste area and, second, tolerant ecotypes that colonize the area contaminated by heavy metals.

Let us look at the questions such a situation should immediately suggest to you!

QUESTION How can we determine whether the plants that occur on waste contaminated by heavy metals are in fact different from normal plants? In other words, do all plants of *A. tenuis* have heavy-metal tolerance throughout the entire range of the species, or are we dealing with distinct ecotypes?

ANSWER When commercially produced seeds of *A. tenuis* are sown on waste contaminated by heavy metals, the majority will germinate, but after several weeks only a few will be left alive. Figure 9 shows the effects of planting 5 000 commercially produced seeds of *A. tenuis* on mine waste contaminated by copper. As you can see, only a few individuals are surviving and, therefore, tolerant—this selection has occurred within 8 weeks of sowing!

Using this approach, Gartside and McNeilly have shown that the frequency of heavy-metal tolerance in commercially produced populations of *A. tenuis* is about 0.08 per cent. It is likely, therefore, that the presence of tolerant individuals on contaminated soils is due to disruptive selection, the majority of plants in contaminated environments being killed off, leaving only a small proportion of tolerant individuals to survive.

QUESTION Of course, another question one can ask is: if ecotypes are restricted to a particular habitat, what is limiting the distribution of the tolerant ecotype from uncontaminated environments? It is quite clear that the distribution of the non-tolerant ecotype is limited to uncontaminated environments because of the toxic nature of the heavy-metal ion. But why is the tolerant ecotype equally restricted to environments contaminated by heavy metals?

ANSWER If we sample plants in a normal pasture close to an area contaminated by heavy metals, very rarely do we find plants that will tolerate heavy metals. Evidence suggests that tolerant ecotypes are at a competitive disadvantage beside non-tolerant plants under normal, uncontaminated, conditions, and would, therefore, be very quickly ousted from such environments.

We have seen that the two ecotypes do, in fact, occupy mutually exclusive environments and, as a result, population divergence has occurred.

The major effect of the toxic ion is the complete inhibition of root elongation in non-tolerant individuals and the reduction of root elongation in tolerant plants. In the 1960s Wilkins capitalized on this response and, by comparing root elongation in de-ionized water with that in de-ionized water containing 7.5 parts/10^6 zinc as zinc sulphate (zinc tolerance) or 12.5 parts/10^6 lead as lead nitrate (lead tolerance) or 0.5 parts/10^6 copper as copper sulphate (copper tolerance), he was able to determine the levels of tolerance of individual genotypes.

Figure 10 shows the rooting response of 4 individuals of *A. tenuis* growing in 7.5 parts/10^6 zinc solution. As you can see, there are two non-tolerant individuals that show complete inhibition of root elongation. The other two show different rates of root elongation and, therefore, differ in their degree of tolerance. In fact, the index of tolerance shows a continuous pattern of variation within tolerant populations, ranging from 10 to 98 per cent tolerance.

At this stage we should point out that tolerance is not universal; a copper-tolerant plant is not necessarily also lead-tolerant, and vice versa. What is very interesting, however, is that where more than one heavy metal occurs in mine waste, the plants have evolved tolerance to all the metals—another striking example of the variation in biological systems.

The evidence so far suggests that tolerant and non-tolerant plants do constitute distinct and divergent ecotypes—but we have said nothing about whether or not the variation in heavy-metal tolerance is genetically determined.

If tolerant plants are taken from an area contaminated by heavy metals and transplanted into uncontaminated soil, their tolerance is retained in cultivation. If these plants are then allowed to flower and set seed as a *polycross* (that is, every plant has an equal opportunity of fertilizing every other plant), their progeny will also be tolerant.

polycross

QUESTION What does this tell us about heavy-metal tolerance?

ANSWER The fact that the variation in heavy-metal tolerance is 'passed on' to the offspring generation tells us that it is a heritable character.

Figure 8 Professor Tony Bradshaw at Trelogan mine, Clwyd.

Figure 9 The selection of copper-tolerant individuals of *A. tenuis* from a normal commercial seed population.

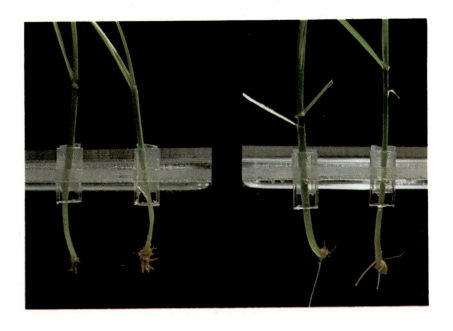

Figure 10 The root response of two non-tolerant individuals (left) and two tolerant individuals (right) of *A. tenuis* in 7.5 parts/10^6 zinc solution.

QUESTION Can you recall from Unit 12 the verbal definition of the narrow-sense heritability?

ANSWER The narrow-sense heritability is the proportion of the phenotypic variance that is additive genetic in origin.

$$h_N^2 = \frac{\text{additive genetic variance}}{\text{phenotypic variance}}$$

QUESTION Can you recall from Unit 12 how h_N^2 may be estimated?

ANSWER The narrow-sense heritability can be estimated from the parent–offspring regression. When the mean of the offspring generation is regressed on to the value of one parent, the regression coefficient, b, is equal to $\frac{1}{2}h_N^2$. If, on the other hand, the mean of the offspring generation is regressed on to the mean of both parents, then $b = h_N^2$.

QUESTION Why are we interested in the narrow-sense heritability of characters?

ANSWER *Take a quick look back at Unit 12, Section 12.5 (p. 573).*

The important point to remember about the narrow-sense heritability and the additive genetic variance is that only the additive effects of the genes are expected to reappear in the progeny generation in such a way as to lead to similarities between the parental and offspring generations.

We can prove this relationship statistically by looking at the covariance of the parents and their offspring, which is equal to $\frac{1}{4}D_R$. Biologically, this can be explained by the fact that dominance relationships in the parental generation will break down at meiosis and be recreated at random in the progeny generation.

Consequently, provided there are no environmental similarities between the two generations, only the additive effects of the genes, that is, the relative proportions of increasing and decreasing alleles, will lead to similarities between parental and offspring generations.

ITQ 4 Pairs of parents of *A. tenuis* were taken at random from a zinc-tolerant population at Trelogan mine in Clwyd and crossed to produce a progeny generation.

The two generations were grown in two separate and randomized experimental layouts in an uncontaminated environment and the tolerance of both parents and offspring was assessed according to the length of the roots.

To make life a little easier, we give you the values for $\sum (x - \bar{x})(y - \bar{y})$ and $\sum (x - \bar{x})^2$, where the x variate is the mean of both parents and the y variate is the mean of 20 offspring in each case.

$$\sum (x - \bar{x})(y - \bar{y}) = 1\,800$$
$$\sum (x - \bar{x})^2 \qquad = 2\,200$$

Calculate the narrow-sense heritability h_N^2 of zinc tolerance in this population.

Of course, until the significance of this estimate has been assessed, we cannot be sure that this relationship has meaning. However, the standard error of the regression coefficient was found to be 0.12 in this case, so the degree of association between the parents and their offspring is statistically significant.

Care must be exercised in extrapolating to the natural environment from data collected from cultivated plants or from plants grown under conditions that are very different from normal. The plant breeder estimates the narrow-sense heritability in the same environment as that in which selection is to be applied. By contrast, many characters of ecological and/or genetic importance are assessed in entirely different environments from those in which selection took place, and is still taking place. For instance, $h_N^2 = 0.82$ for zinc tolerance in this population of *A. tenuis*, but the necessary measurements were taken in an environment very different from that in which the plants normally grow and where h_N^2 may be far less.

However, some idea of the potential for evolution within a population may be assessed from the narrow-sense heritability and estimates of the selection differential, but these are not reliable.

13.3.3 Gene flow

Often the distances separating the extremes of the tolerant and non-tolerant sub-populations may be as small as 100 metres, the boundary between the two being some 0.5–1.0 m only. Recently, a lead-tolerant sub-population of *A. tenuis* was found on an old, lead-contaminated, air shaft, which was surrounded by uncontaminated pasture land and thus a non-tolerant sub-population. The total area of the contaminated habitat was less than 8 m², and the boundary between the two sub-populations was a matter of centimetres.

In Section 13.2.2 we discussed migration and, in particular, gene flow in plants. You will recall that gene flow was very high over short distances (see Fig. 7 on p. 598), falling off to approximately 2 per cent at a distance of 8 metres and maintaining this degree of gene flow over 34 metres.

The boundary between tolerant and non-tolerant sub-populations must, therefore, be subject to a high rate of gene flow in either direction, and the distances separating the extremes of the sub-populations are such that gene flow between the two must occur. Consequently, in order to maintain the integrity of the two populations, divergent selection pressures in the two environments must be operating at a level that counteracts any gene flow.

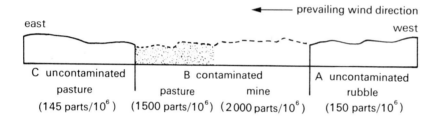

Figure 11 Transect of Drws y Coed mine, North Wales. Underneath are shown copper concentrations in various parts of the transect.

In his investigation of the balance between gene flow and selection, McNeilly concentrated on a small copper mine at Drws y Coed in North Wales, which is situated in a U-shaped valley with an east–west orientation and is subject to a prevailing westerly wind (see Fig. 11).

McNeilly sampled plants upwind from the mine and found a marked transition between the two adjacent sub-populations. Downwind, however, he found a gradual transition, tolerant plants being found some 160 metres away from the boundary between mine and pasture. The difference between the two transects is surprising and cannot be explained by differences in the sharpness of the environmental boundary; the transition from normal conditions to mine conditions, A–B, and from mine to normal conditions, B–C, occurs over the same very short distance.

> QUESTION Can you think of a way to examine gene flow that uses the information we have about the tolerance of the plants in the two transects?
>
> ANSWER McNeilly's method was to compare the tolerance of the adult plants with the mean tolerance of the seeds they produced *in situ*. In the upwind transect, the seeds of the mine plants were slightly less copper-tolerant than the plants that gave rise to them, whereas in the downwind transect, the seeds were more tolerant in the pasture conditions than the plants that gave rise to them.
>
> Clearly, gene flow was acting in a polarized direction, introducing non-tolerant alleles to the tolerant sub-population, that is, A–B, and tolerant alleles to the non-tolerant sub-population, that is, B–C.

In the upwind transect, the establishment of non-tolerant alleles in the tolerant population is severely checked by the intense selection pressures that operate against non-tolerant genotypes in contaminated areas—hence the marked transition between the two environments. In the downwind transect, the balance between gene flow and selection appears to be more in favour of gene flow and it is likely that the tolerant genotypes would not be subjected to intense selection.

Why should this be? The reduced competitive ability of tolerant genotypes severely limits their establishment in normal pastures. In this case, however, the soil is very poor, being typical upland grassland—acid, low in nutrients and productivity; this may explain the more gradual transition.

In the TV programme 'Heavy-metal Tolerance in Plants', we shall continue with this discussion of heavy-metal tolerance and see how our understanding of this subject has led to the reclamation of areas that would otherwise be social and environmental disasters.

You should now attempt SAQ 2 (p. 609).

13.4 Population growth

To end this Unit we shall consider two further 'forms' of selection that may act on a population, namely *r and K selection*. As the parameters *r* and *K* are derived from the equations describing population growth, it is first necessary to consider population growth in some detail.

13.4.1 Exponential growth—non-overlapping generations

Let us assume that the rate of reproduction remains constant over a period—that is, one female leaves two females, two leave four, etc—so that the rate at which the population as a whole increases, is a multiple of the number of organisms already present in the population. A population with x females increases in number x times faster than a population with 1 female, though the individual rate per female is the same.

exponential growth of populations

In many organisms, the various generations do not overlap; that is, by the time the new generation emerges, the old one is dead. This is the pattern for annual plants and many insects.

In a species that reproduces sexually, if each female leaves two females and each male leaves two males (1 : 1 sex ratio), then a population of N individuals will have $N \times 2$ individuals in the next generation. For example, a population of 20 individuals will have 20×2 individuals in the next generation.

Let us formulate this into a simple mathematical expression:

$$N = R_0^t N_0 \tag{1}$$

where N is the new population size

   N_0 is the initial population size

   t is the number of generations

   R_0 is the replacement rate (mean number of progeny left per individual)

For example, a species of plant leaves only seed to survive the winter. The initial population size was 300 individuals, and the replacement rate was found to be 4 (that is, every individual leaves 4 offspring in the next generation). What will the population size be after five generations?

$$N = R_0^t \times N_0$$

where R_0^t is the replacement rate to the power of the number of generations.

Thus,

$$N = 4^5 \times 300$$
$$N = 307\ 200$$

Therefore, the new population size after 5 generations would be 307 200 individuals.

> **ITQ 5** A species of butterfly breeds in late summer and leaves only eggs to survive the winter. One population was observed to increase from 600 butterflies to 1 500 in 1 generation. Predict the population size after 2 further years.

On the basis of this exercise the population size after a further 2 years would be expected to be 9 375 individuals. We have, however, made implicit assumptions about the availability of food and space, the absence of disease, predation and acts of God!, and we have not allowed for variation between individuals. We must, therefore, move on to consider first the effects of the breeding system, and, second, the effects of environmental pressures.

13.4.2 Exponential growth—overlapping generations

In populations in which breeding goes on all the time and generations necessarily overlap, a more general exponential growth equation is given by (see Fig. 12):

$$\frac{dN}{dt} = rN \tag{2}$$

where dN/dt is the rate of change of the number of individuals with respect to a certain time, that is, the *relative* growth rate

where N is the number of individuals in the populations at any given time

t is the time interval (units in hours, days or years)

r is the intrinsic rate of increase (sometimes called the Malthusian parameter), equal to the difference between births and deaths.

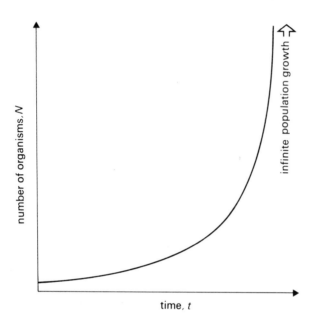

Figure 12 The exponential growth curve, the slope at any point of which, dN/dt, is equal to rN.

Let us consider an example. In an expanding population of lice, r was found to be 0.5 lice per louse per day. What is the rate of population growth per day, dN/dt, in a population of 200 lice?

$$\frac{dN}{dt} = rN = 0.5 \times 200 = 100 \text{ lice}$$

and the new population size would be 300 lice.

> **ITQ 6** In an expanding population of pond weed, r was found to be 0.1 plants per plant per week. (a) What is the rate of population growth per week in a population of 800 plants? (b) What will the new population size be after 1 week?

13.4.3 Logistic growth

Of course, exponential growth will continue only if the various components of the ecosystem can maintain and support the demands of the ever-increasing population for food and space. At some stage the population size will be limited by the availability of essential resources within the ecosystem, and we clearly require an equation that allows for this. In fact, dN/dt averages zero or near zero in most populations, the size of the population fluctuating around some average value. That is, the population has reached equilibrium.

Figure 13 shows the common type of growth curve that expanding populations follow to reach their limit. This limit, which is the number of organisms the environment can just maintain, is often called *the carrying capacity of the environment*, denoted by

logistic growth of populations

carrying capacity of the environment

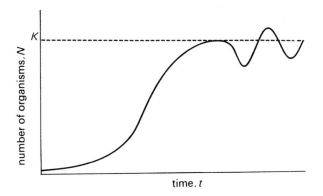

Figure 13 The logistic growth curve, the slope at any point of which, dN/dt, is given by $rn(K - n)/K$.

K. At K, the relative growth rate must on average be equal to zero because the environment cannot maintain any further increase in population size.

Let us look more closely at the logistic growth curve and equation, which is often called the Verhulst–Pearl equation. This takes the form

$$\frac{dN}{dt} = rN\left(\frac{K - N}{K}\right) \tag{3}$$

which you should recall as being the exponential equation, $dN/dt = rN$, but with the added term $(K - N)/K$. This term has been constructed so that when N increases, $(K - N)/K$ decreases and so dN/dt decreases. Where $N = K$, $(K - N)/K = 0$ and $dN/dt = 0$.

However, when N is close to zero, that is, in an expanding population, $(K - N)/K$ is close to 1, and hence dN/dt is very close to being equal to rN (i.e. close to exponential growth).

If N exceeds K, that is, exceeds the resources of the environment, then $(K - N)/K$ becomes negative and dN/dt is also negative, (i.e. there is a negative growth rate) and N will decrease towards K. In this way, the population size will oscillate around the optimum carrying capacity, K, of the environment.

Consider two numerical examples.

1 A population of 500 individuals has an intrinsic rate of increase of 0.03 per year. What is the growth rate of the population per year when the carrying capacity, K, of the environment is equal to 6 000?

$$\frac{dN}{dt} = rN\left(\frac{K - N}{K}\right) = 0.03 \times 500 \times \left(\frac{6\,000 - 500}{6\,000}\right) = 15 \times 0.917 = 13.75$$

Therefore, the population will increase by 13.75 individuals in the year.

2 Keeping the same values for K and r but using a value of 6 500 for N:

$$\frac{dN}{dt} = rN\left(\frac{K - N}{K}\right) = 0.03 \times 6\,500 \times \left(\frac{6\,000 - 6\,500}{6\,000}\right) = 1\,9\dot{5} \times -0.083$$
$$= -16.185$$

Therefore, the population will decrease by 16.185 individuals in the year.

ITQ 7 A species of water-weed has colonized a pond measuring 3×2 metres. Assuming that each individual requires an area of $2\ cm^2$, and that the population size in 1974 was 2 000 individuals with an intrinsic rate of increase of 0.03 per year, what would the population size be in 1975?

Perhaps the greatest drawback to our growth equation is that when the population size is only 1, dN/dt is highest. But it is obvious that in an organism that reproduces sexually, this is just not so. In fact, populations with very small numbers are often in severe trouble and we obviously require a more exact description that includes a level below which the population is doomed to extinction. This level, M, is termed the survival value of the population.

The simplest way to express such a level is to include the parameter M in the exponential growth equation. Thus,

$$\frac{dN}{dt} = rN\left[\left(\frac{K-N}{K}\right)\left(\frac{N-M}{N}\right)\right] \tag{4}$$

In this equation, the growth rate becomes negative if N is less than M, leading to the extinction of the population. If, on the other hand, N is much larger than M, the new term is of minor importance.

Again, let us consider two numerical examples:

1 A population consists of 6 000 individuals and has an intrinsic rate of increase of 0.12 per year. The carrying capacity of the environment is equal to 10 000 individuals, and the value of M is 25. The growth rate of the population is arrived at in the following way.

$$\frac{dN}{dt} = rN \times \left(\frac{K-N}{K}\right) \times \left(\frac{N-M}{N}\right)$$

$$\frac{dN}{dt} = (0.12 \times 6\,000) \times \left(\frac{10\,000 - 6\,000}{10\,000}\right) \times \left(\frac{6\,000 - 25}{6\,000}\right)$$

$$\frac{dN}{dt} = 720 \times 0.4 \times 0.99 = 285.1$$

Therefore, the rate of growth of the population is 285.1 individuals in the first year.

2 Now, consider an example when N is close to M (the survival value). A population consists of 30 individuals and has an intrinsic rate of increase of 0.15 per year. The carrying capacity of the environment is 2 000 individuals and the value of M is 28. The growth rate of the population will be:

$$\frac{dN}{dt} = rN \times \left(\frac{K-N}{K}\right) \times \left(\frac{N-M}{N}\right)$$

$$= (0.15 \times 30) \times \left(\frac{2\,000 - 30}{2\,000}\right) \times \left(\frac{30 - 28}{30}\right)$$

$$= 4.5 \times 0.985 \times 0.07$$

Instead of multiplying this out, we are going to use two separate manoeuvres to illustrate the importance of $(N - M)/N$ when the population is close to the survival value.

(a) Ignoring M, the rate of growth would be $4.5 \times 0.985 = 4.43$ individuals per year, and the new population size would be 34.

(b) Taking M into account, the rate of growth would be $4.5 \times 0.985 \times 0.07 = 0.31$. The population is barely increasing and it is quite likely that it has reached the limit for survival.

> **ITQ 8** A population of herbivores consisting of 300 individuals, and having an intrinsic rate of increase of 0.5 per year, lives in an area of grassland that has a carrying capacity, K, of 305 individuals. The survival value, M, of the population is 2. Calculate the population size after 1 year.

You should now attempt SAQ 3 (p. 610).

13.5 *r* and *K* selection

The two parameters r and K, used in the logistic growth equation

$$\frac{dN}{dt} = rN\left[\left(\frac{K-N}{K}\right)\left(\frac{N-M}{N}\right)\right]$$

are determined ultimately by the genetic constitution of the population, and are thus subject to selection and evolutionary pressures.

Evolutionary geneticists have only recently realized the significance of r and K selection; we are really discussing a frontier subject that in the next few years will almost certainly prove to be most important in terms of understanding the evolution of populations that are adapted to environments that differ in their degree of permanence. The majority of research into r and K selection has been in plant populations, and for this reason we shall concentrate mainly on such populations.

13.5.1 r selection and r strategy

Populations of the annual meadow grass *Poa annua* are frequently found as pioneer colonies on ephemeral habitats such as derelict housing sites, new clearings in forests and mud-banks in rivers.

Such populations are adapted for life in short-lived, unpredictable, habitats and thus require a high intrinsic rate of increase, r, in order to exploit the habitat to its maximum, as quickly as possible, and before it again becomes unsuitable. For example, a population of *Poa annua* colonizes a mud-bank in a river, which has been exposed because of an exceptionally low summer rainfall. The population must reproduce rapidly and set seed that is available for the colonization of other habitats before the mud-bank disappears when the river begins to rise to its natural level during the winter rains. Because of the ephemeral nature of the habitat, the population will always fall within, and never go beyond, the ascending stage of the logistic growth curve, where dN/dt approximates to rN, and genotypes with a high intrinsic rate of increase, r, will be consistently selected for. Genotypes that substitute an ability to compete in crowded conditions (where $dN/dt = 0$ and $N = K$) for a high value of r will not have such a selective advantage.

Populations that depend on a high r value are termed r *strategists*, and the selection for genotypes with a high r value is termed r *selection*.

r strategy
r selection

$$\text{genetic variation in } r \to \frac{\text{selection for maximum } r}{(r \text{ selection})} \to \text{populations with } r \text{ strategies}$$

13.5.2 K selection and K strategy

In contrast, K *strategists* are populations that, because of living in permanent habitats such as climax communities*, are very close to their saturation level, at which $dN/dt = 0$ and $N = K$. In such crowded conditions, a high intrinsic rate of increase is of very little use. Think of the logistic growth equation. Where $N = K$ or is near to K, dN/dt will be equal to or near to zero, because little or no space will be available for seeds to germinate and grow.

K strategy

QUESTION In crowded conditions, can you suggest what particular characters selection will favour?

ANSWER Genotypes with a superior competitive ability will be consistently selected for and this is termed K *selection*.

K selection

K strategists are plants or animals that seize and hold a 'piece' of the habitat and maintain the population at equilibrium in the most dense conditions. Individuals less able to survive and reproduce under long-term competition will be selectively eliminated.

$$\text{genetic variation for } K \to \frac{\text{selection for maximum } K}{(K \text{ selection})} \to \text{populations with } K \text{ strategies}$$

The two forms of selection are not mutually exclusive. r is always subject to some selection, upward or downward, and even during the exponential growth phase, a population must be subject to some degree of K selection. In the main, however, when extreme K selection occurs resulting in stable long-lived individuals, an evolutionary decrease in r usually results.

* A *climax community* is one that has reached the end of its ecological sequence of succession; it is stable and permanent.

607

If we think in terms of strategy in allocation of photoassimilate*, an *r* strategist clearly needs to allocate as much energy to reproduction as possible. As competition is of little or no importance, vegetative growth-form and habit are near to neutral characters in terms of selection. *r* selection will, therefore, favour those genotypes that allocate photoassimilate to reproductive strategy rather than to individual vegetative/competitive strategy; the individuals with the greatest reproductive potential will be consistently selected for.

On the other hand, a *K* strategist needs to allocate as much photoassimilate as possible to individual vegetative strategy in a way that will favour superior competitive ability. As very little space is available for seed germination and growth and there is, therefore, little commitment to reproductive strategy, selection will favour those genotypes with a superior vegetative/competitive strategy. Included in this strategy, however, is vegetative reproduction, for example, stolon formation in grasses. (The stolon of grasses is very similar to the runner on strawberry plants.)

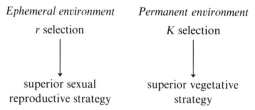

r and *K* strategists can be thought of in terms of a polymorphism—a polymorphism that is initiated by the longevity of the habitat. Such a polymorphism enables a species to exploit different parts of the habitat. The analogy between *r* and *K* strategists and ecotypes is quite obvious. In both *r* and *K* strategists the different ecotypes are genetically and physiologically adapted to different aspects of the habitat, thus allowing a greater intraspecific diversity.

You should now attempt SAQ 4 (p. 610).

Appendix 1   List of symbols introduced in the text

$G \times E$ genotype $\times$ environmental interaction

N_0 the initial population size

R_0 the replacement rate

R_0^t the replacement rate to the power of the number of generations

N the number of individuals in the population at any given time

$\dfrac{dN}{dt}$ the rate of change of the number of individuals with respect to time—the relative growth rate

t the time interval (units in hours, days, etc.); also used for the number of generations

r the intrinsic rate of increase (the Malthusian parameter)

K the carrying capacity of the environment

M the survival rate of the population

* *Photoassimilate*; the food storage compounds (sugars and starches) synthesized during photosynthesis.

Self-assessment questions

SAQ 1

(a) Briefly discuss the basis of continuous variation.

(b) Define the narrow-sense heritability h_N^2.

(c) Given that the additive genetic variance for height in a population of the grass *Agrostis tenuis* is 30 cm^2 and the total phenotypic variation is 50 cm^2, calculate the narrow-sense heritability, h_N^2.

(d) List the three types of selection (excluding r and K selection) that may operate on a continuously varying character.

(e) The mean height of a population of the grass *Lolium perenne* is 20 cm. A plant breeder selects only those individuals above 30 cm high and produces a population with a mean of 35 cm. What type of selection is the plant breeder exercising?

SAQ 2 SAQ 2 has a sequential question and answer structure. *Answer (a); then check your response against the Answers to SAQs (p. 613) before proceeding to (b).*

Excessive concentrations of sodium chloride are toxic and inhibit plant growth. However, in certain areas that are subject to frequent immersion in salt-water, for example, sea cliffs (salt spray), salt marshes (sea-water immersion) and inland salt-water lakes, a limited number of plant species survive and reproduce in spite of what would otherwise be toxic concentrations of salt.

As ecological geneticists, we would be interested in the occurrence of plants in such a hostile environment. For the purpose of this SAQ, we are going to consider the grass *Festuca rubra*, which is growing in a salt marsh habitat that is subject to frequent submersion. Immediately, we would ask, are all individuals of *F. rubra* salt-tolerant, can *F. rubra* grow only in salt-marshes, or are the individuals that do grow in salt-marsh habitats salt-tolerant ecotypes?

(a) How could you test out the three possibilities?

We can postulate, therefore, that the individuals that occur on the salt marsh are specifically adapted, salt-tolerant, ecotypes.

You may have suggested taking known numbers of a commercial sample of seeds of *F. rubra* and growing these on soil contaminated by sea-water. Your answer would be quite correct. The important point is that the majority of seedlings would die very quickly, leaving one or two salt-tolerant individuals surviving. In other words, selection for salt tolerance allows the salt-tolerant individuals within the population to be observed.

Using the root-elongation test, salt tolerance in *F. rubra* can be quantified in a similar manner to that in which we dealt with heavy-metal tolerance. Results obtained from testing many individuals in this manner indicate that salt tolerance is a continuously distributed variate.

We are now interested in determining how much of the variation in salt tolerance is genetically determined and available for selection. 10 individuals were taken from the salt marsh and 2 from the normal pasture that was adjacent; all were tested for salt tolerance by the root-elongation method in 10.0 parts/10^6 sodium chloride solution.

These individuals were grown in compost and pairs of parents were selected at random to give a biparental progeny generation. Seed was collected from each parent. The seeds were grown and 10 progeny from each parent were tested for salt tolerance.

The regression of the mean of the offspring on to one parent was found to be 0.45.

(b) What is the narrow-sense heritability, h_N^2 for salt tolerance in this population of 10 tolerant and 2 non-tolerant plants?

(c) Does this estimate mean anything on its own?

Given that the standard error of the estimate for h_N^2 is equal to 0.04, we can now be sure that 90 per cent of the variation in salt tolerance is additive genetic in origin in this population, when measured in this specific environment.

609

Now, of course, one of the useful properties of the narrow-sense heritability is its capacity to predict outcomes. In Unit 12 we saw that the response to selection, R, was equal to the product of the selection differential, S, and the narrow-sense heritability; that is, $R = h_N^2 \times S$.

Let us say that the mean of the parental generation as a whole is 50 per cent salt tolerance, and we select those individuals with a tolerance of over 60 per cent, the mean of the selected group being 65 per cent tolerance.

(d) Calculate the selection differential.

Now, we know that the narrow-sense heritability is equal to 0.90.

(e) What would be the response to selection in this case?

So, the mean of the progeny generation derived from the selected groups of parents would be 63.5. But what does this mean?

(f) Is it realistic to assume that the response to selection or even the heritability means anything for salt-tolerant populations actually living in salt-contaminated environments?

SAQ 3 A population of elephants consisting of 25 individuals in 1974 has an intrinsic rate of increase of 0.5 per year, and a survival value of 2 individuals.

Each elephant requires 100 kg of vegetation per day and the total productivity per year of the 9 000 km² available for grazing is 2 500 kg per square kilometre.

(a) Calculate the rate of population growth over 1 year and the population size in 1975.

(b) A new population of elephants joins our population and puts the total number of elephants up to 50 individuals. Assuming that all things are equal, calculate the rate of population growth over 1 year.

SAQ 4 Samples of the grass *Poa annua* were taken from two distinct populations (A and B) and grown in cultivation. After allowing the plants to flower, the seeds were collected and again grown in cultivation. Figure 14 shows representative samples of the two populations after 8 weeks of growth in a greenhouse.

Figure 14 Representative plants of *Poa annua* from two different populations (scale in centimetres).

The characteristics of the environments in which the two populations were found are shown below.

population A   permanent grazed pasture in the East Riding of Yorkshire consisting of *Poa annua*, *Agrostis tenuis* and *Lolium perenne*

population B   ephemeral building site on the University campus at Hull

Carefully examine both representative plants, 1 and 2, and in the light of your knowledge of r and K strategists, distinguish (giving reasons) which plant is representative of which habitat. In other words, which is the r and which is the K strategist?

Answers to ITQs

ITQ 1 To answer this question we could do no better than to quote Bumpus. 'It is quite as dangerous to be conspicuously above a certain standard of organic excellence as it is to be conspicuously below the standard.' In other words, selection was favouring the mean expression of the characters that Bumpus was examining, and the extreme phenotypes were eliminated—this is, of course, a classic example of stabilizing selection.

ITQ 2 Your completed frequency-distribution chart should look something like Figure 15. Using this method, it is quite obvious that a morph–ratio cline occurs from

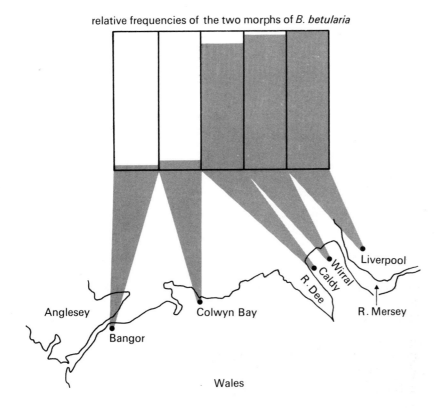

relative frequencies of the two morphs of *B. betularia*

Figure 15 The distribution of the two morphs of *B. betularia*. The shaded portions of the rectangles represent the proportions of the melanic morph.

industrial Merseyside into North Wales. The frequency of the typical morph increases as one goes away from a polluted environment into the relatively unpolluted Welsh countryside.

ITQ 3 (a) An ecotype is a genetically differentiated population or race that is restricted to a specific environment. For example, in a species that has a wide ecological distribution, certain phenotypes will be suited to specific soils: calcifuge ecotypes will occur on acidic soils, calcicole ecotypes will occur on calcaerous soils and salt-tolerant ecotypes will occur in soils exposed to salt (sea) spray, estuaries and salt-marshes.

(b) Allopatric speciation may arise as the result of populations of the same species occupying environments that are isolated as a result of physical barriers. Migration between such populations is restricted and evolutionary pressures within the two populations (genetic drift, mutation and selection) allow them to diverge. When the populations have accumulated too many genetic differences, they will no longer hybridize and can be considered as separate species.

Speciation may also arise in the absence of physical barriers to migration and this is referred to as sympatric speciation. Provided that the balance between migration and divergent selection pressures is in favour of selection, genetic differentiation may arise within the two populations, leading eventually to the formation of new species.

(c) (i) When the two populations are occupying only slightly different environments, but gene flow is at a level that selective forces cannot eliminate, the two populations are likely to converge.

611

(ii) When two mutually exclusive environments occur, the selection pressures within the two populations are likely to be greater than the migration rate and there will be considerable genetic differentiation in the two populations.

ITQ 4

$$b = \frac{\sum (x - \bar{x})(y - \bar{y})}{\sum (x - \bar{x})^2} = \frac{1\ 800}{2\ 200} = 0.82$$

As we are regressing the mean of the offspring generation on to the mean of both parents, $b = h_N^2$. Thus the narrow-sense heritability is equal to 0.82.

ITQ 5 The rate of increase,

$$R_0 = \frac{1\ 500}{600} = 2.5$$

Therefore, the new population size, $N = R_0^t \times N_0$

$$N = 2.5^2 \times 1\ 500 = 6.25 \times 1\ 500$$

$$N = 9\ 375$$

and the new population size would be 9 375 individuals.

ITQ 6

$$\frac{\mathrm{d}N}{\mathrm{d}t} = rN = 0.1 \times 800 = 80 \text{ plants per week}$$

and after 1 week, the new population size would be $800 + 80 = 880$ plants.

ITQ 7 The rate of population growth is given by

$$\frac{\mathrm{d}N}{\mathrm{d}t} = rN\left(\frac{K - N}{K}\right)$$

We know the value of $r = 0.03$ per year

and the value of $N = 2\ 000$

What about K? Well, the area of the pond is $300 \times 200 \text{ cm}^2$ and 1 individual requires 2 cm^2. Therefore, the total number the pond can hold is given by $(300 \times 200)/2$, which is 30 000 individuals. Therefore, $K = 30\ 000$. (We are assuming that K is determined only by the amount of space available.)

$$\frac{\mathrm{d}N}{\mathrm{d}t} = 0.03 \times 2\ 000 \times \left(\frac{30\ 000 - 2\ 000}{30\ 000}\right) = 60.0 \times 0.93 = 55.9$$

$$\frac{\mathrm{d}N}{\mathrm{d}t} = 55.9$$

Therefore, the population size in 1975 was $2\ 000 + 55.9 = 2\ 056$ individuals. (As we cannot have 0.9 of an individual, we round off to the nearest whole number.)

ITQ 8

$$\frac{\mathrm{d}N}{\mathrm{d}t} = rN \times \left(\frac{K - N}{N}\right) \times \left(\frac{N - M}{N}\right)$$

$$= 0.5 \times 300 \times \left(\frac{305 - 300}{305}\right) \times \left(\frac{300 - 2}{300}\right)$$

$$= 150 \times 0.016 \times 0.993$$

$$= 2.38$$

Therefore, the new population size after 1 year should be $300 + 2.38 = 302$ individuals (to the nearest whole number); in this case $\mathrm{d}N/\mathrm{d}t$ is being limited by the carrying capacity, K, of the environment.

Answers to SAQs

SAQ 1 (Objectives 2, 3 and 4)

(a) The multiple-factor hypothesis states that continuous variation is generated as a result of the independent segregation of many genes that affect the same character, the phenotypic expression of a genotype being dependent on the sum total of all the individual cumulative or additive effects of each gene. When the effects of such genes are small in relation to environmentally induced variation, the discontinuity associated with genetic segregation becomes indiscernible in the phenotypic distribution.

(b) The narrow-sense heritability is the proportion of the phenotypic variance that is additive genetic in origin.

$$h_N^2 = \frac{\text{additive genetic variance}}{\text{phenotypic variance}}$$

(c) $h_N^2 = \dfrac{30}{50} = 0.60$ or 60 per cent.

(d) The three types of selection that may operate on a continuously varying character are

(i)   stabilizing selection, in which the mean phenotype is the optimum phenotype

(ii) directional selection, in which the optimum phenotype is not the mean phenotype

(iii) disruptive selection, in which there are two or more optimum phenotypes.

(e) The plant breeder is exercising directional selection as the mean of the population is directed from 20 to 35 cm.

SAQ 2 (Objective 5)

(a) If *F. rubra* can grow only in salt-contaminated areas then we would not expect to find the species in any other habitat. A close look at many lawns, bowling greens and golf courses would confirm that *F. rubra* is not restricted to a salt-contaminated environment.

If individuals of *F. rubra* were taken from the salt marsh, grown in compost in a greenhouse and sprayed regularly with sea-water, most of the plants would survive quite happily, reach maturity and set seed.

Conversely, if individuals of *F. rubra* were taken from a normal pasture and subjected to the same treatment, only a very few, if any, would survive, the majority turning brown and eventually dying.

(b) As the regression coefficient, *b*, of the mean of the offspring against parent is equal to $\frac{1}{2}h_N^2$, then h_N^2 is equal to $0.45 \times 2 = 0.90$. That is, 90 per cent of the variation for salt tolerance in this population, measured in this particular environment, is genetically determined.

(c) No. The statistical significance of this estimate must be assessed by estimating its standard error (see Unit 12, SAQ 5 on p. 580 and its answer on p. 583).

(d) The selection differential is the mean of the selected group of parents expressed as a deviation from the mean of the whole parental group before selection was applied. Thus, $S_d = 65 - 50 = 15$, and the selection differential is equal to 15.

(e)

$$R = h_N^2 \times S$$
$$= 0.90 \times 15$$
$$= 13.5$$

and the response to selection would be equal to 13.5. In other words, the mean of the progeny derived from the selected group of parents will deviate by 13.5 from the mean of the whole parental generation before selection was applied.

The mean of the progeny generation, therefore, will be, $50 + 13.5 = 63.5$.

(f) Care must always be taken in extrapolating to natural situations from results of observations made in cultivation. We know that the narrow-sense heritability is the proportion of phenotypic variance that is additive in origin, that is,

$$h_N^2 = \frac{\frac{1}{2}D_R}{\frac{1}{2}D_R + \frac{1}{4}H_R + E}$$

But, necessarily, this estimate has been made in environmental conditions that are very different from a salt-marsh habitat, and this is quite likely to lead to a heritability estimate that differs from one taken in the wild. Nevertheless, some idea of the possible response to selection can still be obtained.

SAQ 3 (*Objective 6*)

(a) First we need to find the value of K. 1 elephant requires 365 000 kg per year and the environment provides 22 500 000 kg per year. Therefore $K = 22\,500\,000/365\,000 = 616$.

$$\frac{dN}{dt} = rN\left[\left(\frac{K-N}{K}\right)\left(\frac{N-M}{N}\right)\right]$$

$$= 0.5 \times 25 \times \left(\frac{616-25}{616}\right) \times \left(\frac{25-2}{25}\right)$$

$$= 12.5 \times 0.96 \times 0.92$$

$$= 11.0$$

Thus, $dN/dt = 11$, and the new population size after 1 year would be 36 individuals.

(b)

$$\frac{dN}{dt} = rN\left[\left(\frac{K-N}{K}\right)\left(\frac{N-M}{N}\right)\right]$$

$$= 0.5 \times 50 \times \left(\frac{616-50}{616}\right) \times \left(\frac{50-2}{50}\right)$$

$$= 25 \times 0.92 \times 0.96$$

$$= 22.08$$

Therefore the population increases by 22 individuals in the year.

SAQ 4 (*Objective 7*)

Plant number 1 is leafy, has very few flowering tillers, and is clearly allocating a high proportion of photoassimilate to a vegetative strategy. It is, therefore, likely to be a K strategist, and as such will be representative of a population that has evolved under a K selection regime in a permanent and competitive environment. Plant number 1 is, therefore, representative of population A taken from a permanent grazed pasture in the East Riding of Yorkshire.

Plant number 2 has many flowering tillers and less prolific vegetative tillers. Clearly, a high proportion of photoassimilate is being allocated to a reproductive strategy. As such, plant number 2 is an r strategist, and is representative of a population that has evolved under an r selection regime in an ephemeral environment. Plant number 2 is, therefore, representative of population B taken from the ephemeral building site on the University campus at Hull.

It is interesting to note that the seeds were all planted at the same time and the two plants are thus the same age, but if you look very carefully at the photograph, you will observe some signs of senescence in plant number 2, indicating that the K strategist has a longer life-span than the r strategist and is thus able to seize and hold on to its own living space.

Bibliography and references

1 General reading

Ford, E. B. (1975) *Ecological Genetics*, 4th edn, Chapman & Hall.

Head, J. J. (ed.) (1973) *Readings in Genetics and Evolution*, Oxford University Press.

Jones, D. A., and Wilkins, D. A. (1971) *Variation and Adaptation in Plant Species*, Heinemann.

Acknowledgement

Grateful acknowledgement is made to K. Mather for permission to use Figure 2, which has been redrawn from an illustration in *Genetical Structure of Populations*, Chapman & Hall (1973).